Military Enlistment of Hispanic Youth

Obstacles and Opportunities

Beth J. Asch, Christopher Buck, Jacob Alex Klerman,

Meredith Kleykamp, David S. Loughran

Prepared for the Office of the Secretary of Defense

Approved for public release; distribution unlimited

 NATIONAL DEFENSE RESEARCH INSTITUTE

The research described in this report was prepared for the Office of the Secretary of Defense (OSD). The research was conducted in the RAND National Defense Research Institute, a federally funded research and development center sponsored by the OSD, the Joint Staff, the Unified Combatant Commands, the Department of the Navy, the Marine Corps, the defense agencies, and the defense Intelligence Community under Contract W74V8H-06-C-0002.

Library of Congress Cataloging-in-Publication Data

Military enlistment of Hispanic youth : obstacles and opportunities / Beth J. Asch ... [et al.].
    p. cm.
    Includes bibliographical references.
    ISBN 978-0-8330-4572-0 (pbk. : alk. paper)
    1. United States—Armed Forces—Hispanic Americans. 2. Recruiting and enlistment. I. Asch, Beth J.

UB418.H57M55 2009
355.2'236208968073—dc22

                               2009050790

The RAND Corporation is a nonprofit research organization providing objective analysis and effective solutions that address the challenges facing the public and private sectors around the world. RAND's publications do not necessarily reflect the opinions of its research clients and sponsors.

RAND® is a registered trademark.

Published 2009 by the RAND Corporation
1776 Main Street, P.O. Box 2138, Santa Monica, CA 90407-2138
1200 South Hayes Street, Arlington, VA 22202-5050
4570 Fifth Avenue, Suite 600, Pittsburgh, PA 15213-2665
RAND URL: http://www.rand.org/
To order RAND documents or to obtain additional information, contact
Distribution Services: Telephone: (310) 451-7002;
Fax: (310) 451-6915; Email: order@rand.org

Preface

An ongoing concern of Congress, the Department of Defense (DoD), and the armed services is whether the military represents U.S. society at large. An implicit goal is that diversity in the armed services should approximate the diversity of the general population. A key aspect of that diversity is the representation of Hispanics. Furthermore, when military recruiting becomes more challenging, policymakers need to ensure that policies are in place to effectively enlist youth in key demographic groups, such as Hispanics.

Hispanics are a growing segment of the youth population, yet they have historically been underrepresented among military recruits. A widely cited reason is Hispanics' below-average rate of graduation from high school, combined with the services' preference for recruits with high school diplomas. But other, less studied, factors may also contribute. Such factors might include lack of language proficiency as reflected in aptitude test scores; fertility choices; health factors, such as obesity; and involvement in risky activities, such as the use of illegal drugs. These factors, to the extent they are present in the Hispanic population, could adversely affect the services' ability to meet their enlistment standards.

Our project, "Hispanic Youth in the U.S. and the Factors Affecting Their Enlistment," analyzed the factors that lead to the underrepresentation of Hispanic youth among military enlistments. To help policymakers evaluate the feasibility of improving Hispanic enlistments by recruiting more intensively from among the population that is qualified for service and the implications of recruiting Hispanics who are less qualified, we also analyzed both the nonmilitary opportunities available to qualified Hispanic youth and the consequences of recruiting less-qualified Hispanic youth.

This report should be of interest to policymakers and researchers concerned about social representation in the military, opportunities for Hispanic youth, and military recruiting policies.

This research was conducted for the Defense Human Resources Activity and for the Office of Accession Policy, Office of the Under Secretary of Defense for Personnel and Readiness. This study was conducted in the Forces and Policy Resources Center of RAND's National Defense Research Institute, a federally funded research and development center sponsored by the Office of the Secretary of Defense, the Joint Staff, the Unified Combatant Commands, the Department of the Navy, the Marine Corps, the defense agencies, and the defense Intelligence Community.

Comments are welcome and may be addressed to Beth Asch at Beth_Asch@rand.org. For more information on RAND's Forces and Policy Resources Center, contact the Director, James Hosek. He can be reached by email at James_Hosek@rand.org; by phone at 310-393-0411, extension 7183; or by mail at RAND, 1776 Main Street, Santa Monica, California 90407-2138. More information about RAND is available at www.rand.org.

Contents

Figures

Tables

Summary

Hispanics are underrepresented among military recruits. In 2007, Hispanics made up 17.0 percent of the general population (ages 18 to 40) but only 11.4 percent of Army enlistment contracts and 15 percent of Navy enlistment contracts. While the trend is upward (in 1994, 6.6 percent of Army contracts and 8.9 percent of Navy contracts were Hispanic),[1] Hispanics are still underrepresented.

Social representation within the armed forces is an ongoing concern of policymakers. Indeed, each year, the Department of Defense is required by Congress to publish statistics on the social representation of the armed forces in terms of such characteristics as race, ethnicity, marital status, and age. An implicit goal is that diversity in the armed forces should approximate diversity in the general population. Furthermore, recruiting challenges in meeting enlistment goals mean that the services need to understand the factors affecting the supply of key demographic groups, including Hispanics.

The underrepresentation of Hispanics is puzzling, considering that survey data on young people's attitudes toward the military consistently indicate that Hispanic youth are more likely than other groups to express a positive attitude toward the military. For example, in the December 2007 poll of American youth ages 18 to 24 conducted by the Department of Defense, 12.6 percent of Hispanic respondents stated they were probably or definitely going to join the military, compared with 10.1 percent of black respondents and 6.6 percent of white respondents (Defense Human Resources Activity, 2008).

The more positive attitude of Hispanics toward the military would suggest that, all else being equal, Hispanics should be overrepresented, not underrepresented. However, other factors may be at play. Hispanic youth may face greater challenges in meeting one or more of the military's enlistment standards. The services screen applicants in terms of education, aptitude, health, moral character, and other factors. Insofar as Hispanic youth differ from other groups in terms of these factors, they will be disqualified at different rates.

---

[1] The figures on enlistments are based on the authors' computations using Army and Navy enlistment contract data; data on civilian representation are based on the authors' computations using the Current Population Survey. An important caveat when comparing social representation over time is that the definition of racial and ethnic representation changed because of a government-wide change in the standard definitions of race and ethnicity in federal data collections as of January 1, 2003. As a result of this change, agencies—including the Department of Defense—must offer individuals the opportunity to select one or more races when reporting race, and the categories for ethnicity must include "Hispanic or "Latino" and "Not Hispanic or Latino." In this report, the term "Hispanic" is used broadly to encompass those of Hispanic and Latino descent.

The research summarized in this report analyzes the role of the services' entry standards in disqualifying Hispanic youth for military service. For comparison's sake, we also examine qualification rates for white and black youth. The study is designed to answer three key questions:

- Which entry standards are the most likely to disqualify Hispanics from military service, and how does this compare with other groups?
- If recruiting standards were relaxed somewhat, what would be the effect on military performance, using retention and promotion as metrics of performance?
- What actions could be taken to increase Hispanic enlistments? Specifically, to the extent the services recruit more intensively among Hispanics, blacks, and whites who qualify for service, which segments of the qualified market are most likely to find military service attractive—those with higher aptitude, better education, or fewer qualifications?

Data Sources and Limitations of the Analysis

To study disqualification with respect to entry standards, we analyzed the characteristics of the general population with respect to the services' major entry standards. We used two nationally representative datasets: the National Longitudinal Survey of Youth (NLSY) from 1997 to 2003 and the National Health Interview Survey (NHIS) from 1998 to 2001. Both data sources provide information on demographic and other individual characteristics pertinent to entry standards used by the services.

To examine the effects of relaxing standards on military performance, we consider the downstream performance of military entrants who vary in terms of their quality and other characteristics. Some of these entrants received waivers of the enlistment standards (e.g., the service may permit the enlistment of individuals who have disqualifying characteristics). Performance is measured in terms of the retention and promotion outcomes of recruits. For our analyses, we created a longitudinal data file of military careers by merging annual master file and transaction records for all enlisted personnel with enlistment records for individuals entering service between fiscal years 1988 to 2003.

However, our analyses have four limitations. First, the information in the two national datasets differs in some areas, and the reasons for these differences are unclear. For example, comparison with other data sources suggests that the NLSY overstates high school graduation among Hispanics. Second, the services' standards are, in some cases, highly detailed. For example, in the case of health standards, the standards specify a degree of severity or a time component (e.g., the standards require that the most recent instance of a disqualifying condition such as asthma have occurred in childhood). Large, nationally representative datasets do not provide information at this level of detail. Third, the datasets may not provide information on the specific disqualifying factor. For example, some of the military's health and moral fitness standards are not captured in the national datasets. Finally, at the discretion of the individual service, some standards may be waived, and the waiver process is not always clearly defined. Thus, the estimates of the percentage of the population who are disqualified does not account for the waiver process and the percentage who might qualify after receiving a waiver.

We also note that the term "Hispanic" encompasses a highly diverse population in terms of country of origin, geographic region, and immigrant status, to name a few characteristics (Tienda and Mitchell, 2006). However, most of our data sources do not provide enough detail to allow analyses of subgroups. Thus, for the most part, the analysis considers all individuals of Hispanic and Latino descent as "Hispanic," recognizing that this is a broad categorization.

Factors That Disqualify Potential Hispanic Recruits

Analysis of the NLSY data reveals that a relatively small percentage of youth, regardless of race or ethnicity, would qualify for military enlistment. Figures S.1 and S.2 show the cumulative effect of key enlistment standards in the areas of education (high school diploma or General Education Degree), aptitude (Armed Forces Qualification Test score, [AFQT]), weight, number of dependents, convictions, and drug-related offenses. Results are shown by race/ethnicity for males and females, respectively, by service. Only 46 percent of white males, 32 percent of black males, and 35 percent of Hispanic males would be eligible to enlist in the Marine Corps, the service with the cumulatively least stringent enlistment standards. For females, the corresponding figures are even lower: 35 percent for white females, 22 percent for black females, and 24 percent for Hispanic females.

Major Disqualifying Factors
We found that the major characteristics that disproportionately disqualify Hispanic youth are lack of a high school diploma, lower AFQT scores, and being overweight. Each is briefly discussed below.

Figure S.1
Cumulative Percentage of Males Passing Standards for Education, AFQT, Weight, Dependents, Convictions, and Drugs, by Service

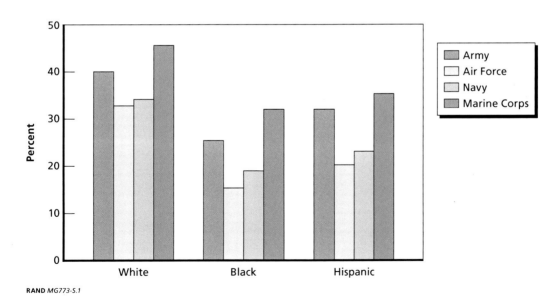

RAND MG773-S.1

Figure S.2
Cumulative Percentage of Females Passing Standards for Education, AFQT, Weight, Dependents, Convictions, and Drugs, by Service

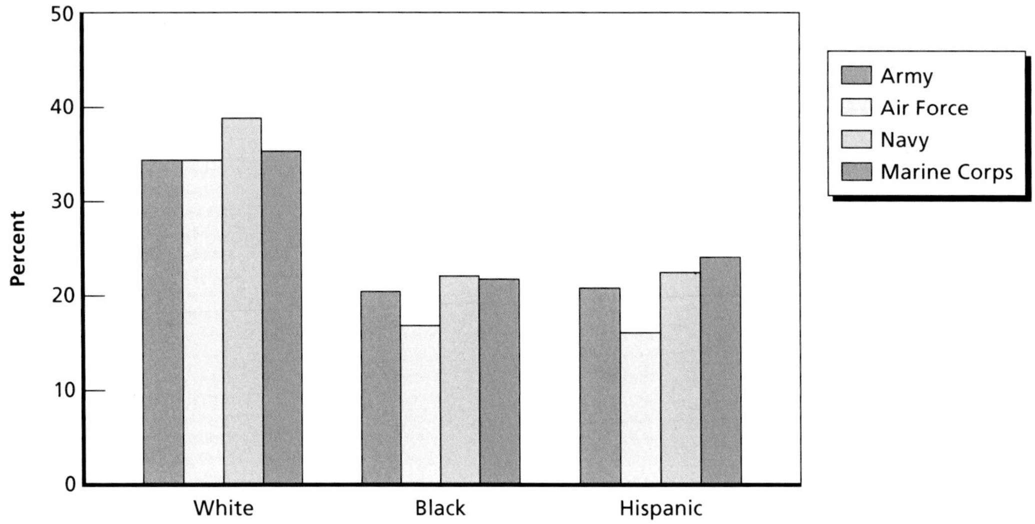

SOURCES: NLSY 1997; authors' computation.
RAND *MG773-S.2*

Since the services prefer high school graduates, Hispanics' lower high school graduation rate goes a long way toward explaining why they are underrepresented among enlistments. In the NLSY sample, 74 percent of Hispanic males are high school graduates, compared with 85 percent of white males. (As noted above, the NLSY may overstate high school graduation among Hispanics; the actual graduation rate may be lower.)

Though important, education is not the only major disqualifying characteristic of Hispanic youth. Hispanics who are high school graduates often fail to meet other enlistment standards. The services require that potential recruits take the AFQT. Based on their test results, potential recruits are placed in one of five categories (Category I is the highest). The services strongly prefer recruits whose score places them in Category IIIB or higher. The Department of Defense (DoD) restricts the annual accession of those in Category IV (the next-to-lowest category) to 4 percent of the total, and prohibits all recruiting from Category V (the lowest category). Only 36 percent of young Hispanic high school graduates would score in AFQT Category IIIB or above, compared with 68 percent of white high school graduates. A key implication of this result is that increasing the high school graduation rate among Hispanic youth may not lead to comparable increases in enlistment eligibility.

As is well known, childhood and adult obesity has increased among the U.S. population. This trend has important implications for military recruitment: fewer youth are likely to meet the services' weight standard. Comparing Hispanics with other groups, we see that weight is another important disqualifying characteristic. Hispanics are considerably heavier than others: on average, Hispanic males weigh almost ten pounds more than white males. Seventy-nine to 91 percent of white males meet the service weight standards (weight standards vary by service), compared with only 71 to 88 percent of Hispanic males. Among females, the percentage who meet the weight standards is even lower; 63 to 82 percent of white females meet the standards, compared with only 49 to 71 percent of Hispanic females.

Other Disqualifying Factors

In addition to education, AFQT score, and weight, the military also evaluates recruits in terms of other factors, including major and minor medical conditions, number of dependents, and moral character (recent drug or alcohol use or engaging in illegal activities). These factors have a less important effect on Hispanic recruitment, as described below.

We evaluated health in terms of three factors: weight (discussed above), and "major" and "minor" disqualifying conditions. We termed as major health conditions those that are non-waiverable; they include blindness, hearing problems, and organ failures, such as stroke and hypertension. Minor conditions are those that might be waived at the discretion of the individual service; they include such conditions as hay fever and attention deficit disorder.

Our research shows that Hispanics have a lower prevalence of disqualifying major and minor conditions than whites. That is, except for weight, Hispanics tend to be healthier than whites. Research suggests that better-than-expected health in the Hispanic population may be due to the large proportion of immigrants; immigrants in general, regardless of ethnicity, tend to be healthier than the native-born U.S. population.

However, Hispanics are more likely to be disqualified because of weight. On balance, taking all three health standards together (weight, major conditions, and minor conditions) Hispanic males are disqualified at about the same rate as whites. Hispanic females are substantially more likely to be overweight than white females, and more likely to be disqualified.

Our analysis indicated that number of dependents is another disqualifying characteristic for Hispanics. Though not as important as weight, education, or AFQT, it is a significant factor, especially for females. Twenty percent of young Hispanics (ages 17 to 21) have children, compared with only 9 percent of whites.

The final set of qualification characteristics considered are those related to moral character. The NLSY queries respondents about drug and alcohol use and engagement in past illegal activities, but problems with these data mean that these questions are unlikely to provide an accurate picture of the extent to which Hispanics will be disqualified relative to other groups due to moral character. Since misdemeanors can be waived and individuals would presumably curtail their drug use in advance of taking a drug test, it is unclear how misdemeanors and drug use would affect eligibility rates.

Actions That Could Improve Hispanic Enlistments

The military could increase Hispanic representation by increasing the pool of qualified individuals, by relaxing recruiting standards, or by recruiting more intensively from among those who are already qualified. Our analysis evaluated some of the implications of these potential courses of action.

Increasing the Pool of Qualified Individuals

One approach to increasing the pool of qualified individuals is to implement policies that encourage high school graduation and improved educational achievement (resulting in improved AFQT scores). However, the practicality of such policies is questionable. An important factor explaining Hispanics' relatively low graduation rates and AFQT scores is family background, such as mother's education and family income. Thus, without addressing underlying family and economic circumstances, the role of the services may be limited. Yet, even at the margin, the services may be able to encourage some students to graduate who otherwise might not. Summarizing the academic literature, Heckman (1995) notes that motivation plays an important role in economic achievement. Since many Hispanic youth are favorably inclined toward military service, it is possible that appropriate motivating factors could increase young Hispanics' graduation rate and educational achievement.

Relaxing Recruiting Standards

Recruiting more intensively from the pool of qualified Hispanics will be challenging. Most likely, increasing representation among the Hispanic population will involve enlisting more marginal recruits. The services already have programs that seek to identify the best of these marginal recruits or to improve the AFQT, weight, or educational outcomes of those recruits. These programs are not specifically targeted to improve Hispanic representation, but insofar as Hispanics are more likely to be disqualified because of AFQT, weight, and lack of a high school diploma, these programs are more likely to increase Hispanic enlistment.

A key question is whether the programs that increase the enlistment of somewhat lower quality Hispanics will have a large adverse effect on subsequent military performance. Our study provides some information on this question using five metrics of performance: retention at three months (roughly corresponding to completion of boot camp and initial skill training); retention at four years of service (approximately the end of the first term of service); retention at six years of service (approximately the end of the first term of service in technical skills in the Air Force and Navy); achievement of promotion to pay grade E-5 by four years of service (corresponding to early promotion and perhaps being on the fast track); achievement of promotion to pay grade E-5 by six years of service. The analysis focuses on how varying weight, AFQT, and education affects outcomes, because these are the three main characteristics that disqualify Hispanic youth from service.

Hispanics consistently have higher retention and faster promotion speeds than their white counterparts. For example, the predicted four-year retention rate for Hispanic recruits in the Army is 54 percent—6 percentage points higher than the 48 percent rate predicted for white recruits. (An exception is the Navy, where white recruits are promoted faster.) Blacks also tend to have higher retention and faster promotion, but adjusting for observed characteristics (such as AFQT scores) shrinks the effects of race on outcomes.

As found in past studies, our analysis indicates that higher-quality recruits tend to stay longer and be promoted faster. In the case of AFQT, the positive relationship between AFQT and retention is strongest for white recruits. Those who fail to complete high school have lower retention relative to high school graduates, regardless of race or ethnicity. Those who are overweight, especially 20 pounds overweight relative to the service standard, have lower retention than those within five pounds of the standard. Overweight recruits have poorer outcomes, regardless of race and ethnicity. For example, Hispanics in the Navy who are more than 20 pounds overweight have a promotion rate of 19 percent to E-5 within six years, compared with a promotion rate of 23 percent for Hispanics who are within five pounds of the Navy's weight standard.

While the effect of standards on military career outcomes is significant, the effects of race and ethnicity are even larger. Lower-quality Hispanics compare well with higher-quality white recruits who have similar observed characteristics. For example, the four-year retention rate of white recruits in AFQT Categories I and II, adjusted for observable characteristics, is predicted to be 42 percent, while the adjusted four-year retention rate for similar Hispanic recruits in AFQT Categories IIIB and IV is predicted to be 50 percent. For blacks, the four-year predicted retention rate in AFQT Categories IIIB and IV is predicted to be 45 percent. Consequently, lower-quality minorities are more likely to remain in service and be promoted than higher-quality white recruits. Presumably the better outcomes are attributable to minorities being better matched to the military in terms of factors unobserved to the analysts (such as higher motivation or better opportunities in the military).

The implication of this analysis is that targeting the recruitment of more marginal Hispanic recruits is not likely to have adverse effects on retention or promotion speed. In fact, the analysis suggests that, at the margin, faced with the decision to recruit minorities over identical white recruits, the services would gain more person-years, via greater retention, by favoring minorities because of greater retention. In the case of Hispanics, greater retention is an additional way to improve Hispanic representation. As Hispanics stay longer, their relative representation in the enlisted force increases. The implication discussed above is based on analytical results only and does not take into account the DoD's obligation to avoid racial discrimination. Furthermore, we note that untargeted recruitment of marginal recruits would hurt retention and promotion rates.

Recruiting More Intensively from Among Qualified Individuals

To understand better how the military might meet the career aspirations of qualified Hispanics and improve the supply, our study described the career and schooling choices of young people, by race/ethnicity, as they transition from adolescence to young adulthood. These choices represent the competition; in other words, they are the external opportunities that the military must overcome to compete successfully for qualified youth.

To analyze education and career choices, we divided potential recruits into three groups: (1) the *least qualified*, defined as those without a high school diploma or who scored in AFQT Category IV or V; (2) the *next-most qualified*, defined as those who are high school graduates in AFQT Categories IIIA and IIIB; and (3) the *most qualified*, defined as those who are high school graduates in AFQT Categories I and II.

The analysis suggests that improving Hispanic enlistments within each group will be challenging. A large percentage of Hispanic youth (55 percent) are in the least qualified group. In addition to having less education, this group tends to have much poorer labor market outcomes relative to high school graduates. For example, at age 22, the median civilian wages of Hispanics in this group are about 20 percent less than the wages of Hispanic high school graduates. Not only are their employment rates and wages lower than high school graduates, they also tend to fail other enlistment criteria, such as weight standards and recent drug use. It is unlikely that this group would perform well in the military.

The military already disproportionately recruits those in Group 2, relative to their representation in the population. Specifically, 73 percent of young Hispanic recruits are in Category IIIA and IIIB (this group represents only 33 percent of the general Hispanic population). Increasing Hispanic representation by recruiting more intensively from this group will mean pulling from a population that is already heavily recruited.

However, it may be possible to make additional recruiting inroads with this group by recognizing that they have strong interest in college. Over 60 percent of the total people in this group are enrolled in either two- or four-year college. Hispanic enrollment patterns differ only somewhat from that of other groups in that Hispanics are much more likely to attend two-year college and more likely than either white or black youth to receive a training certificate or vocational license. Strong interest in two-year college may reflect lack of resources for education. To the extent that even two-year college involves considerable expense in terms of forgone earnings for those who do not work full-time, the high college attendance rates among this group suggest individuals might be responsive to the suite of educational benefits the military offers.

Group 2 youth are also strongly attached to the labor market, with around 80 percent of 20-year-olds employed. Working while in school is common, especially among Hispanics and whites.

The final group is high school graduates in AFQT Categories I and II. For Hispanics, this group is quite small. These individuals also have excellent college and career opportunities. College attendance, especially at four-year colleges, is relatively high. Specifically, 52 percent of Hispanics in the NLSY sample in this group attended four-year colleges compared with 67 percent of white youth and 63 percent of black youth.

This group also has excellent employment opportunities. Employment rates increase with age, and by age 23, about 80 percent of youth in AFQT Categories I and II are employed. Their average earnings are higher than the earnings offered by the military.

Educational and career outcomes for white, black, and Hispanic youth in this group are similar. However, military application rates are lower among Hispanics in this group when compared with white and black youth. Attracting recruits from this group into the military will require focusing on the availability of educational benefits, the leadership opportunities and scope of responsibilities offered by the military, the opportunities to serve one's country, and other nonpecuniary benefits of service.

Policy Implications

The analyses yield several policy insights and implications, for both the current recruiting environment and for social representation of the military in the long term. With respect to the current environment, the Army did not meet its recruiting mission in 2005, and has struggled to meet its mission in 2006 and 2007.

A major disqualifying factor for Hispanics is weight. Other services might consider adopting the Marine Corps approach to weight. This service has the most relaxed weight standard in the armed forces, but simultaneously requires applicants to pass a strength test. Adopting this approach throughout the armed services might increase the pool of potential recruits and Hispanic supply, with only minimal effects on attrition, retention, and promotion.

As part of its effort to increase enlistments, the Army has recruited more lower-quality enlistees, relaxed its enlistment standards, and begun several experimental programs to allow applicants who failed to meet standards to qualify for enlistment.

A disproportionate percentage of the lower-aptitude Army recruits are black or Hispanic. We found that lower-aptitude minorities have better retention than higher-aptitude white recruits, all else being equal. An implication of our analysis is that the armed services, while avoiding overt discrimination, should develop recruiting incentives attractive to Hispanics and blacks.

In the longer term, the analysis suggests that identifying and targeting the most motivated of the least-qualified group of Hispanics is a good approach, and is consistent with current efforts like the Army's Tier Two Attrition Screen program. For the most-qualified group, the military must find ways to compete with excellent civilian opportunities. This will include emphasizing the nonpecuniary benefits of service, such as leadership opportunities, higher span of responsibility, and opportunities to serve one's country. Finally, for applicants in the middle range, college seems quite important, especially two-year college for Hispanic youth. Since many do not complete college, and many work while in college, more exploration is needed as to whether these individuals lack resources or have lower educational expectations. In either case, military service as part of one's educational path, along with the suite of educational benefits available to those who serve, could be marketed more heavily to this group.

Educational benefits are only one of many recruiting resources. Little is known about how the supply of Hispanics and black recruits responds to other recruiting resources, such as enlistment bonuses, as well as to external factors including the Iraq war, the civilian economy, and college opportunities. Such information would be useful for developing policy options to increase the supply of Hispanic and black recruits.

Acknowledgments

We thank Curt Gilroy, the Director of Accession Policy in the Office of Secretary of Defense, for his support and guidance as the sponsor of this research effort. We are indebted to John Warner and Curtis Simon at Clemson University for providing data on military accessions and career outcomes, and to Richard Moreno at the Defense Manpower Data Center for assisting us with Military Entrance Processing Station data. We benefited from comments from numerous sources, including Aline Quester and Anita Hattiangadi at the Center for Naval Analyses, and Altagracia Ramos and Naomi Verdugo and their colleagues in the Army's advertising and marketing division. We would like to express particular gratitude to our RAND colleagues Larry Hanser, Nelson Lim, and Elaine Quester, for comments on our interim report, and Ashlesha Datar and Benjamin Karney for comments on the final report. We also wish to thank Christine DeMartini, Roald Euller, and Bogdan Savych for their assistance. We are grateful to Dr. Jane Arabian in the Accession Policy Directorate and Andrea Zucker in the Defense Human Resource Activity, Joint Advertising and Market Research Studies Division, for their comments and assistance.

Abbreviations

AFQT	Armed Forces Qualification Test
ALCPT	American Language Course Placement Test
AR	arithmetic reasoning
ASVAB	Armed Services Vocational Aptitude Battery
AVF	All-Volunteer Force
DMDC	Defense Manpower Data Center
DoD	Department of Defense
ECLT	English Comprehension Level Test
ENTNAC	Entrance National Agency Check
ESL	English as a second language
FAST	Fundamental Applied Skills Training
FLRI	Foreign Language Recruiting Initiative
FY	fiscal year
GED	General Educational Development (test)
HumRRO	Human Resources Research Organization
HSB	High School and Beyond Survey
IRR	Individual Ready Reserve
JAMRS	Joint Advertising Market Research and Studies
MAW	maximum allowable weight
MEPS	Military Entrance Processing Station
MK	Mathematics Knowledge
MOS	Military Operational Specialty
NELS	National Educational Longitudinal Study
NHIS	National Health Interview Survey

NLSY	National Longitudinal Survey of Youth
PC	paragraph comprehension
RMC	regular military compensation
TABE	Test of Adult Basic Education
USMEPCOM	U.S. Military Entrance Processing Station Command
VE	verbal expression
WK	word knowledge

CHAPTER ONE

Introduction: Hispanic Enlistments in Perspective

The representation of Hispanics and other minorities in the military is an ongoing concern of policymakers in Congress and in the military. At the beginning of the All-Volunteer Force (AVF), in 1974, the Senate Armed Services Committee mandated that the Department of Defense (DoD) publish statistics annually on the social representation of the armed forces in terms of such characteristics as race, ethnicity, marital status, and age (Senate Committee on Armed Services, 1974). This concern grew out of the social representation of the draft force during the Vietnam War and criticism that the burden of military service, and especially war casualties, were falling on minorities.

Although DoD did not begin keeping statistics on Hispanics specifically until the early 1970s,[1] evidence based on the experience of blacks indicated that blacks were more likely to serve in combat-related assignments and were more likely to suffer casualties during the Vietnam conflict than whites (Badillo and Curry, 1976; Gimbel and Booth, 1996).[2] The concern about representation did not end with the end of the draft and the beginning of the AVF. While black enlistments were 14.1 percent of Army enlistments in 1971 and 14.2 percent of all DoD enlistments (Binkin and Eitelberg, 1982), they had risen dramatically to 23.2 percent for the Army and 22.8 percent for all DoD by 1975, the second full year of the AVF. By 1979, black enlistments had reached 36.1 percent of Army enlistments and 36.7 percent of DoD enlistments. DoD data reveal that Hispanic enlistments as a share of Army non-prior-service enlistments rose from 6.1 percent in 1973 to 7.5 percent in 1979 and from 5.4 percent to 6.3 percent across all DoD enlistments (Department of Defense, 1997). Although the report of the President's Commission on the All-Volunteer Force (also known as the Gates Commission) predicted that movement to the AVF would not alter the social representation of the armed forces, early experience with the AVF indicated that, indeed, a higher percentage of minorities was serving in the military.[3] By 1980, critics of the AVF pointed to the low quality of all Army recruits of the late-1970s. Concerns about a "hollow Army" led to a 25 percent increase in military basic pay in the early 1980s. As recruit quality increased in the 1980s, Hispanic and black representation changed as well. By 1990, 25.2 percent of Army accessions were black, and 6.2 percent were Hispanic (Department of Defense, 1997), compared with 13.9 percent and 10.8

---

[1]   An overview of the history of Hispanics' contribution to the armed forces since the civil war is given in Rosenfeld and Culbertson (1992).

[2]   Analysis of the qualifying characteristics of Vietnam-era veterans by Gimbel and Booth (1996) reveals that once the authors had controlled for AFQT, a key factor related to occupational assignment, there was no statistical difference in the probability of combat assignment for blacks versus whites.

[3]   An excellent discussion of the history of blacks in the military as the U.S. switched from conscription to the AVF is given in Binkin and Eitelberg (1982).

percent of the 18- to 24-year-old civilian population for blacks and Hispanics, respectively. Thus, blacks were overrepresented among Army recruits and among military recruits in general, while Hispanics were underrepresented relative to the civilian population.

The U.S. Hispanic population has grown. This population differs from other ethnic groups in that it is younger, has a common language, has a relatively low education level on average, includes a sizable percentage of immigrants, and contains an important segment that is not legally documented (Tienda and Mitchell, 2006). These characteristics have different implications for the ability of Hispanic youth to meet the military's entry standards and ultimately for their representation in the military. Because the military has virtually no lateral entry and recruits all individuals at the lowest ranks, the military's success in recruiting and retaining Hispanic enlistees determines the overall supply of Hispanics in the armed services.

Aside from social representation concerns, another reason for interest in the supply of Hispanic recruits is the recruiting challenges experienced recently, primarily by the Army. The Army missed its 2005 recruiting mission and has struggled to achieve its 2006 and 2007 mission, substantially increasing recruiting such resources as enlistment bonuses and recruiters. One part of an effective strategy for meeting these challenges is increasing the supply of youth from growing segments of the youth population, specifically Hispanics. Thus, increasing the supply of Hispanic youth to the military is not just an issue of representation but also a strategy for improving recruiting effectiveness.

Before proceeding, we note that the term "Hispanic" encompasses a highly diverse population in terms of country of origin, geographic region, and immigrant generation, to name a few characteristics (Tienda and Mitchell, 2006). Among the young adult population, the target population for recruiting, about 75 percent of the population is Mexican, about 10 percent is Puerto Rican, and the others come from various other locations (authors' computations). Most of our analysis uses data sources that have insufficient sample sizes to permit disaggregation by subgroup. Thus, for the most part, we group all individuals of Hispanic and Latino descent as "Hispanic," recognizing that this is a broad categorization.

Since 1994, the share of Hispanics among Army enlistments has grown, although recently it has declined. The share nearly doubled from 6.6 percent in 1994 to 12.6 percent in 2003, as shown in Figure 1.1, but has since declined to 11.4 percent.[4] In the Navy, Hispanic representation among Navy enlistment contracts increased from 8.9 percent in 1994 to 15 percent in 2003. Nonetheless, Hispanic recruits are still underrepresented compared with the percentage of 18- to 40-year-old Hispanics among the civilian population as shown in the monthly Current Population Survey data. For comparison's sake, Figure 1.2 shows the percentage of enlistment contracts and the civilian population that is black.

This underrepresentation occurs despite the higher positive propensity of Hispanic youth to enlist. Positive propensity is defined as the percentage of individuals who respond either "probably" or "definitely" to the following question in the semi-annual Department of Defense youth poll: "How likely is it that you will be serving in the military in the next few years?" In the December 2007 poll, 12.6 percent of Hispanic males stated a positive propensity, compared with 10.1 percent of black males and 6.6 percent of white males (Defense Human Resources

---

[4] It should be noted that a governmentwide change in the standard definitions of race and ethnicity occurred in federal data collections as of January 1, 2003. Individuals must now be offered the opportunity to select one or more races when reporting race, and the categories for ethnicity must include "Hispanic or Latino" and "Not Hispanic or Latino."

Figure 1.1
Percentage of Enlistment Contracts and Civilian Population That Is Hispanic

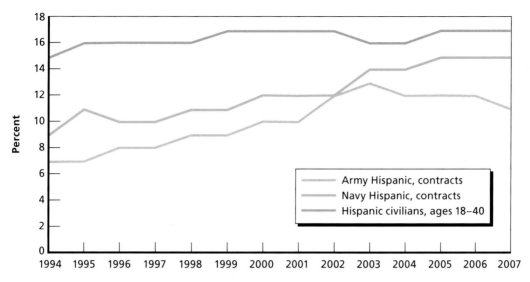

SOURCES: Enlistments: Department of Defense (2007); U.S. Census Bureau (2007).
RAND *MG773-1.1*

Figure 1.2
Percentage of Enlistment Contracts and Civilian Population That Is Black

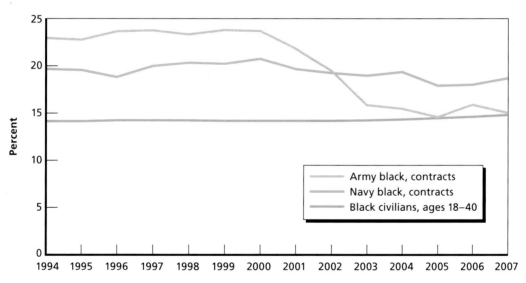

SOURCES: Enlistments: Department of Defense (2007); U.S. Census Bureau (2007).
RAND *MG773-1.2*

Activity, 2008). Past research has documented that among respondents, as well as among high school graduates and seniors, Hispanics are the most likely to state a positive propensity for service (Wilson et al., 2000). The strong positive propensity of Hispanic high school seniors and graduates in the face of underrepresentation suggests that Hispanic graduates and seniors are disqualified for service on the basis of other factors. As will be shown in Chapter Three, a relatively large share of the Hispanic high school graduate civilian young adult population is concentrated near the bottom of the distribution of Armed Forces Qualification Test (AFQT) scores, i.e., AFQT Categories IV and V. These are groups the military severely restricts in terms of its entry standards.[5]

Given the interest of policymakers in social representation and the supply of Hispanic youth to the military, an important policy question is what can be done to improve Hispanic recruitment and representation in the armed forces. This report is designed to answer three key questions:

- Which entry standards are most likely to disqualify Hispanics from military service and how do disqualification characteristics compare with other groups?
- If recruiting standards were relaxed somewhat, what would be the effect on military performance?
- What actions could be taken to increase Hispanic enlistments?

Enlistment Standards

In 2007, the Department of Defense enlisted 177,000 individuals. To ensure that these individuals were qualified for the rigors of military service, the services applied a range of standards. The individual services can impose more restrictive standards than the Department of Defense (DoD) standard. Chapter Two provides a summary of the standards imposed by the individual service branches. A more detailed listing of the qualifications is provided in Appendix A, and supporting material about qualifications is provided in Appendix B.

We based our analyses on the enlistment standards detailed in Appendixes A and B. However, it should be noted that the services change their enlistment standards as recruiting conditions change. For example, as recruiting has become more challenging recently, the Army has relaxed its maximum age standard, allowing individuals up to age 42 to enlist. The information provided in this chapter reflects the standards in effect at the time we gathered the information for the analyses, specifically 2004 and early 2005.

In brief, enlistment standards can be divided into the following categories:

- age
- citizenship
- dependency status
- financial screening
- education
- aptitude
- moral character
- substance use

---

[5]   The AFQT is a composite of the scores from the verbal and mathematic tests from the Armed Services Vocational Aptitude Battery (ASVAB), the ten-part test used to screen all enlisted applicants. The ASVAB is normed against a nationally represented sample of American youth ages 18–23. The AFQT score is expressed on a percentile range and grouped into categories. Category I is the highest, and Category V is the lowest. Category III is divided at the 50th percentile. The percentile divisions are Category I (93–100); Category II (63–92); Category IIIA (50–64); Category IIIB (31–49); Category IV (10–30); Category V (1–9). See Sellman (2004) for discussion of how enlistment standards, particularly aptitude standards, are determined.

- language skills
- homosexual conduct
- height and weight standards
- strength requirements
- medical screening.

These standards are described in greater detail in Chapter Two.
Although applicants who fail to meet enlistment standards are disqualified, they may regain eligibility by obtaining a waiver. Waivers may be issued for most violations; there are only a limited number of disqualifications that cannot be waived. See Chapter Two for more detail on waivers.

Organization of the Report

The remainder of this study focuses on the three key research questions. Chapter Two provides background on the enlistment standards of each of the service branches. The information in this chapter is used to study the extent to which standards screen out youth. Furthermore, because the issue of enlistment standards and waivers is of general interest to the military recruiting policy arena, this chapter provides considerable detailed information on standards and waivers. Readers with less interest in specific standards can skip to Chapter Three.

Chapters Three and Four address our first research question: Which entry standards disqualify Hispanics more than other groups? Chapter Three analyzes the factors in general that disqualify Hispanic youth from military service. Given the large number of health-related enlistment standards, Chapter Four focuses on health standards—specifically, weight requirements and major and minor medical conditions.

Chapter Five addresses the second research question: If recruiting standards were relaxed, what would be the effect on military performance? This chapter presents analyses of the military career outcomes, and the effects of varying entry standards on outcomes for Hispanic recruits relative to other recruits. It also provides information relevant to extending the careers of Hispanic recruits.

Chapter Six addresses the third question: What actions could be taken to increase Hispanic enlistments? One approach to increasing Hispanic enlistments is to increase the number of Hispanic youth who are eligible and would meet the military's entry standards. A second approach is to increase interest and recruit more intensively among those Hispanic youth who already meet these standards, i.e., the qualified Hispanic population. A third approach would be to target recruiting toward less-qualified Hispanics, including granting more waivers to those who qualify for service on the basis of all but one standard. This chapter investigates all three possible approaches. In addition, because recruiting more intensively among the qualified Hispanic population requires information about their competing nonmilitary opportunities, this chapter analyzes work and educational choices available to Hispanics as they transition out of high school and eventually settle into employment in the labor market as young adults.

Finally, Chapter Seven discusses the policy implications of this research.

CHAPTER TWO

Overview of Service Enlistment Standards

In fiscal year (FY) 2007, DoD enlisted 181,170 individuals. To ensure that these individuals were qualified for the rigors of military service, the services applied a range of standards. This chapter provides an overview of the qualifications for enlisting in active duty service established by DoD. Since individual service branches can impose more restrictive standards, this chapter also provides a summary of those standards. A more detailed listing of the qualifications is provided in Appendix A, and supporting material about qualifications is provided in Appendix B. The chapter provides input to the analyses presented in later chapters of the report. Because of the general policy interest in enlistment standards and waivers, considerable detail is provided in this chapter and in the appendixes. Readers can skip this chapter without losing the main points of the report.

The standards summarized in this chapter were used to answer one of the study's key research questions: Which standards disproportionately disqualify Hispanic youth? However, it should be noted that the services change their enlistment standards as recruiting conditions change. For example, as recruiting has become more challenging in recent years, the Army has relaxed its maximum age standard, allowing individuals up to age 42 to enlist. The information provided in this chapter reflects the standards as they were in effect at the time we gathered the information for the analyses (in 2004 and early 2005) and was drawn from the documents listed in Table 2.1.

Waivers

It is important to note that, although applicants who fail to meet enlistment standards are disqualified, disqualified individuals may regain eligibility if a waiver can be obtained. Waivers may be issued for most failures to meet the standards; there are only a limited number of standards that cannot be waived, as seen in Table 2.2. Depending on the type and nature of a waiver, varying levels of authority are required to authorize it. Those waivers that are judged to be more serious are sent up the chain of command. Like the standards themselves, each service branch creates its own policies regarding waivers and their use. Therefore, each branch has different policies. Appendix B provides more details on service policies regarding waivers.

Table 2.1
Enlistment Documents, by Source

Source	Document
DoD	Department of Defense, *Qualification Standards for Enlistment, Appointment, and Induction,* Directive Number 1304.26, March 4, 1994.
Army	Department of the Army, *Regular Army and Army Reserve Enlistment Program,* Army Regulation (AR) 601-210, March 28, 1995.
	Department of the Army, *Standards of Medical Fitness,* AR 40-501, September 30, 2002.
Air Force	U.S. Air Force Air Education and Training Command, *Recruiting Procedures for the Air Force,* AETC Instruction 36-2002, March 18, 2003.
Navy	United States Navy Recruiting Command, *Navy Recruiting Manual–Enlisted,* COMNAVCRUITCOMINST 1130.8F, July 3, 2003.
Marine Corps	Headquarters U.S. Marine Corps, Military *Personnel Procurement Manual,* Volume 2, *Enlisted Procurement,* Marine Corps Order (MCO) P1100.72B, December 10, 1997.
	Headquarters U.S. Marine Corps, *Military Personnel Procurement Manual,* Volume 2, *Enlisted Procurement,* MCO P1100.72C, February 10, 2004.

Enlistment Standards

The purpose of enlistment standards is to ensure that the services select the most qualified recruits. Enlistment standards can be divided into the following categories:

- age
- citizenship
- dependency status
- financial screening
- education
- aptitude
- moral character
- substance use
- language skills
- homosexual conduct
- height and weight standards
- strength requirements
- medical screening.

We discuss each category in turn.

Age

Enlistees must be between ages 17 (with parents consent) and 35, according to the 2005 DoD standard. This requirement ensures that those who enlist are eligible for retirement (which

Table 2.2
Disqualifying and Nonwaiverable Enlistment Standards, by Service

Category	Criteria	Service Army	Air Force	Navy	Marine Corps
Age	Individuals not between 17 and 35 years of age	x	x	x	x
Citizenship	Individuals who are not U.S. citizens or lawful permanent residents of the U.S., or citizens of the Federated States of Micronesia or the Republic of the Marshall Islands	x	x	x	x
Dependents	Single parents	x	x	x	x
	Those with three or more illegitimate children				x
Aptitude	Individuals who do not meet AFQT requirements based on service and education level[a]	x	x	x	x
Moral eligibility	Individuals convicted of drug trafficking, sales, or distribution	x	x	x	x
	Individuals on probation, parole, or civil confinement	x	x	x	x
	Those with pending judicial proceedings or criminal charges against them	x	x	x	x
	Those convicted of serious criminal misconduct while in the Delayed Entry Program (DEP)				x
Substance use	Individuals under the influence during processing or who test positive for drugs or alcohol at MEPS[b]	x	x	x	x
	Individuals with a history of alcohol dependency	x			x
	Individuals with a history of drug dependency	x			x
	Individuals with driving convictions involving drugs other than cannabis, steroids, and prescription drugs				x
	Those with a court conviction for any drug offence (except simple possession of cannabis [30 grams or less] and steroids)				x
	Those who used LSD within two years prior to enlistment			x	
Homosexuality	Homosexual conduct[c]	x	x	x	x
Medical	A variety of medical conditions	x	x	x	x

[a] The only known exception is for the Arabic Language program. For more information see Language Proficiency later in this chapter.

[b] MEPS is the military entrance processing station.

[c] As defined in DoD Directive 1304.26.

requires 20 years of service) by the age of 55. Beyond this requirement, each service branch has set its own maximum age for enlistment, as shown in Table 2.3.

Citizenship[1]

To be eligible for enlistment an individual must be one of the following:

- a U.S. citizen
- an alien lawfully admitted to the United States for permanent residence[2]
- a citizen of the Federated States of Micronesia or the Republic of the Marshall Islands.

These are DoD standards. The Army, Air Force, and Navy allow the enlistment of noncitizen U.S. nationals (see definition in Appendix A). The Marine Corps accepts only noncitizens who "establish a bona fide residence, and establish a home of record in the United States" (MCO P1100.72C, p. 3-25).

While citizenship is not a requirement for enlistment, noncitizens are restricted in their ability to serve in several ways.[3] First, noncitizens have restricted job opportunities in the U.S. armed forces. They are not eligible for appointment as commissioned or warrant officers, because citizenship is a requirement for these positions. Moreover, they are not eligible for security clearance status, which disqualifies them from a variety of jobs. Second, some noncitizens are not allowed to reenlist. The Air Force does not allow noncitizens to reenlist unless they have become citizens during their first term of duty. Similarly, the Army bars "aliens who will have in excess of 8 years of Federal military service (excluding the DEP after 1 January 1985) at the expiration of the period for which they are seeking to reenlist" from reenlistment (AR 601-280, p. 7). The Navy and Marine Corps do not have such restrictions and allow all noncitizens to reenlist regardless of whether or not they have become citizens while serving.

The Marine Corps also specifies that noncitizens who have had residency in countries considered hostile toward U.S. interests "after his/her 15th birthday, [or taken] more than

Table 2.3
Minimum and Maximum Age for Enlisting, by Service

Service	Minimum	Maximum
Army	17	34
Air Force	17	27
Navy	17	28
Marine Corps	17	34

---

[1] Definitions regarding residency and nationality are relevant for the purposes of enlistment. These definitions are provided in Appendix A.

[2] This includes aliens lawfully admitted to the United States as conditional permanent residents. Because conditional permanent residents are technically defined as permanent residents under the law, even though they have an additional condition added to their status (see definitions in Appendix A), they qualify under DoD Directive 1304.26. The Air Force is the only service that does not accept conditional permanent residents. They are disqualified by the following statement: "any applicants who have expiration dates 2 years or less from date of issue on their INS Form I-551 . . . are ineligible for processing or enlistment" (p. 37).

[3] In most cases, U.S. noncitizen nationals are treated as citizens.

two trips to one or more hostile countries within five years proceeding his/her enlistment, excluding school trips, family vacation, sporting events, or other similar, short-lived group sponsored events," require a waiver (MCO P1100.72C, p. 3-31). Latin American countries included on this list are Cuba and Nicaragua.

Number of Dependents

DoD bars applicants who are married with more than two children and nonmarried applicants with any children (Table 2.4). However, DoD allows the services to grant waivers to promising applicants. The Army and Air Force standards are almost identical to the DoD standard. They allow the enlistment of married applicants who have three dependents (a spouse and two additional dependents) before a waiver is required. The Marine Corps and Navy set more restrictive standards, requiring a waiver for married applicants with more than one dependent. Because the spouse of a married applicant is defined as a dependent, applicants enlisting in the Marine Corps or Navy who have any dependents other than their spouses require a waiver. All the services prohibit the enlistment of single parents with any children under the age of 18. This requirement has no exceptions, and waivers are not available for single parents.

The services differ in how they classify dependents. The details of these differences are described in Appendix A.

Financial Screening

All applicants are required to discuss their financial history with service personnel prior to enlistment. The services' policy for financial screening can be found in Table 2.5.

Table 2.4
Number of Dependents Allowed, by Service

Service	Not Married	Married
Army	0	3
Air Force	0	3
Navy	0	1
Marine Corps	0	1

Table 2.5
Financial Screening Policy, by Service

Service	Policy
Army	All applicants must fill out DA Form 3072-2, Applicant's Monthly Financial Statement
Air Force	If an applicant requires a dependency waiver, has a history of financial problems, is over the age of 23, or has ever been married, a financial review is required
Navy	All applicants must fill out NAVCRUIT 1130/13, Enlistee Financial Statement
Marine Corps	A financial review is required for all married applicants and all applicants who indicate they have someone dependent upon them for financial support

Applicants with a history of bankruptcy or bad credit, or who may not be able to meet their current financial obligations, may be disqualified or may be required to obtain a waiver to enlist. The services' financial screening policies are shown in Table 2.6.

Education

In 1987, the Department of Defense implemented a three-tier system to classify the educational credentials of recruits. These tiers are based on past attrition levels and were designed to limit the number of applicants enlisted each year from traditionally high attrition categories (attrition is traditionally highest among those without a high school diploma).

Tier 1. This is the highest educational category. The following candidates are categorized as Tier 1 recruits:

- Traditional high school diploma graduates
- College/postsecondary education students who have attended and successfully completed 15 semester hours or 22 quarter hours of college, regardless of high school or grammar school education
- Alternative or continuation high school graduates who have had the same daytime course and graduation requirements as graduates of the traditional local public school system. Those alternative or continuation graduates who do not meet this description are classified as Tier 2
- Adult high school graduates who have attended and completed an adult education diploma program that included attendance comparable to that of traditional high schools. Adult diploma holders who do not have comparable attendance are classified as Tier 2.

Tier 2. Recruits with the following educational credential are considered Tier 2:

- Applicants who possess a General Educational Development (GED) or other equivalency certificate or diploma
- Applicants who possess an attendance-based certificate or diploma based on course completion rather than a test such as the GED
- Alternative or continuation high school graduates who do not qualify for Tier 1
- Home-schooled individuals who earned a diploma or certificate upon completion of correspondence school coursework
- Correspondence school diploma holders who earned a diploma or certificate upon completion of correspondence school coursework

Table 2.6
Financial Screening Rules of Each Service

Service	Rule
Army	No specific policy
Air Force	40% rule: Applicants must be able to cover their current debt with 40% of their salary
Navy	Applicants must not have debt that exceeds 50% of their annual salary or, if indebtedness includes a long-term mortgage, it must not exceed 2.5 times their salary
Marine Corps	"Applicants will not be enlisted if it appears that they are unable to meet current and expected financial responsibilities with service pay" (MCO P1100.72C, p. 3-34)

- Occupational Program Certificate holders who attended a vocational/technical or proprietary school for at least 675 classroom hours and possess a certificate of attendance or completion indicating such attendance. Correspondence schools offering vocational certificates are not included.

Tier 3. Tier 3 recruits are those who do not have Tier 1 or 2 credentials.

Aptitude

Aptitude testing has been part of applicant screening since World War I (Sellman, 2004). It helps determine eligibility for enlistment as well as qualification for specific jobs. The Armed Services Vocational Aptitude Battery (ASVAB) is a ten-part test that measures verbal, mathematics, and science/technical skills and knowledge. The services combine different ASVAB subtests to form composite scores, and these scores are used to assign recruits to occupations. The AFQT score is a composite of the verbal (word knowledge and paragraph comprehension) and mathematics (arithmetic reasoning and mathematics knowledge) abilities. The AFQT is the primary enlistment aptitude screen used by the services.

The ASVAB is normed against a nationally representative sample of youth, ages 18 to 23, who were administered the ASVAB as part of the National Longitudinal Survey of Youth (NLSY), conducted by the U.S. Bureau of Labor Statistics. This norming permits the scores of military applicants and recruits to be compared with those of the youth civilian population. The AFQT scores are expressed as percentiles and are grouped into five categories (see Appendix A). DoD disqualifies applicants who score in percentiles 1–9 on the AFQT and limits the number of enlistees who score in percentiles 10–30 to 20 percent of the total number of enlistees in any given fiscal year.[4] The services have different score requirements, as seen in Table 2.7.

For Tier I (primarily high school diploma graduates), the Army requires a minimum AFQT score of 16, while the Air Force requires a minimum of 40. These are minimum scores: The services would prefer higher scores if possible. By law, those who are in Tier III, i.e., have not graduated from high school, and who are in AFQT percentiles 30 and below are not eligible for enlistment.

Table 2.7
AFQT Score Requirements, by Tier Level and Service

Tier	Army	Air Force	Navy	Marine Corps
Tier I	16	40	31	21
Tier II	31	50	50	31
Tier III	31	50	50	50

---

[4] The only recruits not required to meet these standards are those enrolled in the Army's Individual Ready Reserve (IRR) Direct Arab Linguist Program. See "English Language Training" in the Language Proficiency section for more information.

Moral Character

Moral character standards are based on past illegal behavior and substance use. To judge the moral character of applicants, each branch of the Armed Forces has created categories of crimes and has set a limit on the number of violations in each category before applicants need waivers or are disqualified. Some of these standards are summarized in Table 2.8; all are described in further detail in Appendix A.

To ensure that all violations are included when assessing the moral eligibility of applicants, an Entrance National Agency Check (ENTNAC) is conducted on all applicants at the MEPS. This check also includes offenses that have been expunged or cleared from a person's record. Each applicant is informed about the depth of the criminal record check used during the enlistment process and is required to disclose all offenses, including those that were expunged or cleared. Applicants who fail to disclose any information regarding their criminal record may be disqualified from enlistment, according to service regulations.

Any applicants with court charges filed or pending against them (including both criminal and juvenile charges), and any applicants under civil restraint (including confinement, parole, and probation) are disqualified and not eligible for waivers. Also, anyone convicted of a felony is ineligible for enlistment. However, the services can grant waivers to this policy.

The above description summarizes DoD policy. Appendix A summarizes the individual services' standards.

Substance Use

Although crimes involving drugs and alcohol are addressed in the moral qualification section, each service has additional requirements specifically targeted at behavior involving controlled substances. Table 2.9 summarizes drug- and alcohol-related activities requiring a waiver by the services. Appendix A provides more details about service drug and alcohol standards.

All the services disqualify applicants who test positive for drugs or alcohol at the MEPS. However, the services have slightly different policies regarding whether or not these applicants can regain eligibility. The Army, Navy, and Marine Corps allow the applicant to retest after a waiting period, which varies by service. The Air Force permanently disqualifies those who fail the drug and alcohol test.

Table 2.8
Illegal Activities That Require Waivers, by Service

Offense Category[a]	Army	Air Force	Navy	Marine Corps
Minor traffic	6 or more	6 or more in any 365 day period during the past 3 years	6 or more	5 or more
Minor non-traffic	3 or more	2 or more in the last 3 years or 3 or more in a lifetime	3, 4, or 5	2 or more for serious traffic; 4 or more class 1; 2–9 class 2
Misdemeanors	2, 3, or 4	1 or more	1, 2, or 3	1, 2, 3, 4, or 5
Felonies	1 or more	1 or more	Not more than 1	1 or more

[a] These categories vary by service and are grouped as shown only for comparison.

Table 2.9
Drug- and Alcohol-Related Activities Requiring Waivers, by Service

Service	Offense			
	Preservice Use of Marijuana	**Preservice Use of Drugs Other Than Marijuana**	**Drug or Alcohol Dependence**	**Drug Trafficking**
Army	Applicants must disclose use, but waivers are not required	Applicants must disclose use, but waivers are not required	Not waiverable	Not waiverable
Air Force	5 or more uses requires a waiver	Use of amphetamines, barbiturates, over-the-counter drugs, or anabolic androgenic steroids is waiverable	Waivers are available for alcohol abuse if applicant has abstained for a minimum of 2 years prior to enlistment; waivers are also available if an applicant is or was involved in a rehabilitation program for the use or abuse of marijuana	Not waiverable
Navy	Waiver not required for experimental or casual use	Use of stimulant or depressant drugs, narcotics, hallucinogenic, or psychedelic drugs is waiverable if not used in year preceding enlistment, LSD use in 2 years prior to enlistment is not waiverable	Prior psychological or physical dependence upon any drug or alcohol is waiverable	Not waiverable
Marine Corps	1 or more use requires a waiver	1 or more use requires a waiver	Not waiverable	Not waiverable

Each service has a different policy on marijuana use. The Army states that applicants having a history of chronic cannabis (marijuana) use or psychological cannabis dependence are disqualified and not eligible for a waiver. The Air Force requires waivers for applicants who have used marijuana more than five times prior to enlistment. The Navy does not require a waiver for experimental or casual use of marijuana. The Marine Corps has a more restrictive policy and requires waivers for all applicants who have ever used marijuana.

Language Proficiency
All the armed services use the AFQT to screen recruits for language skills because the ASVAB, upon which the AFQT is based, is administered in English. Recruits who lack the English skills necessary to qualify based on their AFQT scores are disqualified. The only recruits who are accepted for enlistment with disqualifying AFQT scores are those enlisted in the Army's IRR [Individual Ready Reserve] Direct Arabic Linguist Program, which began on August 11, 2003. This is one of three programs run by the Army to train successfully enlisted recruits in the English language. Besides the Army, the Navy is the only other service providing English language training for enlisted recruits. For more information on these programs, see Appendix A.

Beyond testing the language skills of recruits through the AFQT, each service provides additional instructions as follows:

Army. The Army attempts to identify all recruits who meet AFQT standards but have difficulties speaking or understanding English.[5] Once identified, these recruits are given the English Comprehension Level Test (ECLT). Those who score 69 or below on the ECLT are required to attend English language training.

Air Force. The Air Force also uses the ECLT, but only at the MEPS in San Juan, Puerto Rico, and only when an applicant needs to retake the ASVAB and his or her score indicates a possible "comprehension disability." The instructions provided by the Air Force are as follows:

> The ECLT or ALCPT [American Language Course Placement Test] should be administered to applicants before ASVAB retesting when the original scores indicate a possible comprehension disability. A score of 70 verifies the requirement that an applicant is able to read, write, speak, and understand the English language. (U.S. Air Force Air Education and Training Command, 2003, p. 46).

Navy. Beyond looking at the overall AFQT score of an applicant, the Navy assesses the language capabilities of recruits by examining the Verbal Expression (VE) section of the ASVAB. Any applicant who scores below 43 on the VE section (equated with an 8th grade education) is required to take the Test of Adult Basic Education (TABE). If recruits score below a 34 on the TABE (also equated with an 8th grade education) they are enrolled in the Fundamental Applied Skills Training (FAST) program. This program provides successfully enlisted recruits with either literacy training or English language training, depending on their needs.[6]

Marine Corps. The only mention of English language ability in MCO P1100.72B states that MEPS personnel must certify "that the applicant, without further instruction, is able to read, write, and speak the English language sufficiently to complete recruit training" (MCO P1100.72B, p. 3-30).

Homosexual Conduct

DoD Directive 1304.26 gives a very detailed account of homosexual eligibility requirements. Because this topic is covered in this directive, the individual services do not add requirements to the DoD standards but instead refer directly to those standards. The details of the directive are provided in Appendix A.

Height and Weight Standards

All the services require that applicants be in good physical condition, including specific height restrictions (Table 2.10) and weight standards for given heights. Tables 2.11 and 2.12 give examples of different heights, and the minimum and maximum weights required by the services. The services differ somewhat in how they implement their height and weight standards. The Army breaks down weight standards by height and age; the Air Force and Navy base their weight standards on height alone.[7] The Marine Corps uses age and height-specific standards.

---

5   The process used to identify these recruits is not described.

6   For more information on these programs see Appendix A.

7   The Army's age categories are 17–20, 21–27, 28–39, and 40+ for both men and women. The Marine Corps' age categories for men are 16–20, 21–30, and 31–35; for women, they are 16–20, 21–24, 25–30, and 31–35.

Table 2.10
Maximum and Minimum Heights, by Service and Gender (inches)

	Males		Females	
Service	Minimum	Maximum	Minimum	Maximum
Army	60	80	58	80
Air Force	60	80	58	80
Navy	60	78	58	78
Marine Corps	58	78	58	72

Table 2.11
Examples of Maximum Weight, by Service, Height, and Gender

		Maximum Weight, by Service (lbs)				
					Marine Corps	
Gender	Height (inches)	Army[a]	Air Force	Navy	Shipping	DEP[a]
Male	72	200–210	205	201	213	227–233
Male	68	179–187	184	181	190	203–208
Female	68	154–167	164	170	164–170	
Female	64	137–148	146	156	146–152	

[a] Depending on age.

Table 2.12
Examples of Minimum Weight, by Service, Height, and Gender

		Minimum Weight, by Service (lbs)			
Gender	Height (inches)	Army	Air Force	Navy	Marine Corps
Male	72	131	131	118	131
Male	68	115	115	104	115
Female	68	112	114	101	125
Female	64	102	103	92	110

However, for males, the Marine Corps uses only height-specific (not height- and age-specific) standards at the time individuals "ship" or leave for boot camp. That is, the Marine Corps has a different set of standards for male recruits who ship to recruit training and those who enlist into the Delayed Entry Program. The shipping standards are more stringent and all male Marine Corps recruits must meet them before they enter recruit training. However, much heavier applicants are allowed into enlist into DEP, as seen in Table 2.11.

The Army, Navy, and Air Force have very similar maximum weight requirements for men; the Marine Corps allows slightly heavier applicants to ship to recruit training and much heavier applicants to enlist into DEP, as seen in Table 2.10. The Army, Air Force, and Marine Corps all have the same standards for male recruits' minimum weight while the Navy allows lighter applicants to enlist, as seen in Table 2.12.

In the case of women, the Navy has slightly higher maximum weight standards and lower minimum weight standards than the other services, as seen in Tables 2.11 and 2.12. Additionally, the Marine Corps has higher minimum weight standards, as seen in Table 2.12.

Applicants who exceed the maximum allowed weight are given a body-fat measurement to determine if they exceed the allowable body-fat standards, as shown in Table 2.13. Despite having the most lenient weight standards, the Marine Corps has the most restrictive body-fat requirements.

For complete height and weight charts see Appendix A.

Strength Requirements

The Marine Corps is the only service that has a strength requirement. Before shipping to recruit training, male recruits must be able to do two pull-ups, 35 sit-ups in two minutes, and run 2.5 miles in 13:30 minutes. Female recruits must be able to do a flexed arm hang for 12 seconds, do 35 sit-ups in two minutes and run one mile in 10:30 minutes MCO P1100.72B, p. 3-105). The Army has begun testing the use a strength test, based on the Harvard Step Test.

Medical Screening

DoD places the responsibility for medical screening on the MEPS. Applicants are screened to ensure that they are

- free of contagious or infectious diseases
- free of medical conditions or physical defects that would require excessive time lost from duty or would likely result in separation from the service for medical unfitness
- medically capable of satisfactorily completing required training
- medically adaptable to the military environment
- medically capable of performing duties without aggravation of existing physical defects or medical conditions.

According to U.S. Military Entrance Processing Station Command (USMEPCOM) Regulation 40-1 (DoD, 2002), the medical standards for enlistment prescribed by DoD are the same for all services, and are contained in AR 40-501 (see Table 2.1). However, DoD gives the services the authority to grant waivers in individual cases.

Two types of conditions are disqualifying. The first type of condition is generally not waived. These conditions tend to be major medical issues that would prevent recruits from operating effectively in the field; they include functional limitations, severe hearing and/or vision problems, ailments that require special equipment, heart problems and diabetes. The second type of condition is composed of medical issues that may be waived and may simply

Table 2.13
Allowable Body Fat Measurements, by Gender and Service

Gender	Army	Air Force	Navy	Marine Corps
Male	24–30	20–32	22	18
Female	30–36	20–32	30	26

require additional examination or documentation before a decision is made. Examples of potentially waiverable conditions include allergies to bee stings, back trouble, bone or joint surgery, or a stomach ulcer in the past five years. Appendix A provides a full list of waiverable and nonwaiverable medical conditions.

CHAPTER THREE

Enlistment Qualifications

In Chapter One, we learned that Hispanics are underrepresented among military accessions. Historically, the military services have pointed to relatively low rates of high school completion among Hispanics as the principal reason for this underrepresentation, but a number of other factors may also be at play. This chapter focuses on the importance of five enlistment standards in limiting the pool of eligible Hispanics. We consider the effect on eligibility of these five enlistment standards, compliance with which we can measure reasonably well in survey data. The standards are (1) the Armed Forces Qualifying Test (AFQT) score, (2) body weight, (3) number of dependents, (4) criminal activity, and (5) substance abuse. While these five enlistment standards are not exhaustive of all enlistment standards described in Chapter Two, they are the most significant in determining eligibility in broad segments of the youth population, as we discuss in this chapter. The most significant enlistment standards we do not consider here relate to health conditions other than body weight (i.e., major and minor medical conditions).

Chapter Four examines compliance with health standards using data more suited to that particular domain. Other enlistment standards not examined here relate to sexual orientation, financial condition, strength (Marine Corps only), citizenship, and language proficiency. Citizenship and language proficiency are certainly relevant to the Hispanic population, but they are unlikely to be important by themselves. Citizenship is actually not a requirement for enlistment, although being a legal, permanent resident is.

Data

The data for our analysis come primarily from the 1997 National Longitudinal Survey of Youth (NLSY97), a nationally representative sample of youth ages 12–16 in 1996 surveyed each year between 1997 and 2003.[1] For comparison purposes, some analyses employ an earlier cohort of the NLSY, the NLSY79, a nationally representative sample of youth ages 14–22 when first surveyed in 1979.[2]

We imposed a number of restrictions on both of these samples. These restrictions are detailed in Table 3.1. First, we restricted our samples to youth ages 17 to 21 in 2001 (in the case

---

[1] By design, the NLSY97 oversamples blacks and Hispanics. However, the NLSY97 sample is nationally representative when properly weighted. All tabulations and other analyses employ sample weights. See *NLSY97 User's Guide* (2005) for more about the NLSY97.

[2] For more information about the NLSY97 and NLSY79, refer to http://www.bls.gov/nls/.

Table 3.1
Sample Selection

Sample Restriction	Data Source			
	NLSY79	NLSY97	2003 Youth Poll	2000 Census
Ages 17–21	7,724	7,705	2,374	195,167
Not attending high school	5,894	5,501	1,354	130,657
Not missing data on enlistment criteria	5,497	4,322	1,335	NA

of the NLSY97) and in 1983 (in the case of the NLSY79). We then dropped all individuals who were currently attending high school. We dropped these individuals so that our reference population did not include individuals who had not yet completed high school, but were likely to in the near future (e.g., high school juniors). Finally, we dropped individuals who were missing data necessary to compute compliance with the AFQT, weight, and dependents standards. About 20 percent and 4 percent of the age-eligible NLSY97 and NLSY79 samples are missing data on AFQT. Our final samples are composed of 4,322 and 5,497 youth from the NLSY97 and NLSY79, respectively.

We relied on the 2003 Youth Poll to examine compliance with standards related to drug use and criminal behavior. The 2003 Youth Poll is a nationally representative survey of youth ages 16–21 in 2003. It was designed by the Joint Advertising and Marketing Research & and Studies (JAMRS) Office within DoD to study factors affecting youth perceptions of the military and their propensity to enlist.[3] As explained further below, the Youth Poll included a number of questions on drug use and criminal behavior that are more directly relevant to enlistment standards than the questions on drug use and engagement in past illegal activities found in the NLSY. As with the NLSY samples, we limited our sample from the Youth Poll to youth ages 17–21 who were currently not attending high school, which left us with a sample of 1,335 youth.

Finally, we employed data from the 2000 Census to examine the impact of high school graduation on eligibility in the overall Hispanic population—which, we have discovered, differs significantly from the population of Hispanics surveyed in the NLSY97. We also employed data from the 5 percent Public Use Microsample (PUMS) of the 2000 Census, restricting our sample to youth ages 17–21 who were not currently attending high school ($n = 130,657$).[4]

All statistics reported below are weighted using survey weights provided by each survey.

Caveats

The reader should be aware of three significant limitations of the statistics reported in this chapter. First, the data we used to compute eligibility with enlistment standards do not perfectly align with actual standards in all cases. Our data for computing high school graduation, AFQT, weight, and dependents corresponds reasonably well with service-specific enlistment standards. However, it is much more difficult to determine with our data whether a given

---

3   Refer to Boehmer and Zucker (2004) for information on the 2003 youth poll.

4   Refer to http://www.census.gov/Press-Release/www/2003/PUMS5.html for information on the 2000 Census.

individual would meet the services' standards regarding substance abuse and criminal activity. Second, the individual services can and do issue waivers to certain youth who fail to pass a given enlistment eligibility standard. Thus, our estimates will underestimate the percentage of youth who are in fact eligible to enlist, because we cannot account for the percentage of technically ineligible youth who subsequently receive a waiver. This is especially true in the case of drug use and criminal activity. Finally, as noted in the section above, the NLSY97 appears to have oversampled Hispanic high school graduates. While many of our analyses are limited to high school graduates, it is conceivable that the NLSY97 population of Hispanic high school graduates is not representative of the overall U.S. population of Hispanic high school graduates in ways that are correlated with eligibility on other enlistment standards.

Organization

The remainder of this chapter is organized as follows. We first use data from the 2000 Census to show that Hispanics are underrepresented among military accessions both in the overall youth population and in the population of high school graduates. In the same subsection, we highlight important differences between the Census and NLSY Hispanic samples with respect to high school graduation and discuss what those differences imply for the remaining analyses. The next subsection examines how AFQT standards affect enlistment eligibility. We next examine weight and standards for dependents. Finally, we look at how relaxing particular standards affects overall eligibility, recognizing that eligibility is correlated across specific standards (e.g., individuals who fail to meet AFQT standards are more likely to fail to meet weight standard than are individuals who do meet AFQT standards).

Hispanics in the Census and NLSY97

In this subsection, we make three points regarding Hispanics, high school graduation rates, and military service. The first point is that failure to graduate from high school is clearly a major reason why Hispanics are underrepresented in the military. The second point is that the NLSY97 substantially overestimates high school graduation rates in the general Hispanic population. The NLSY97 Hispanic subsample appears to be unrepresentative in this respect, so we must be careful in how we draw inferences from this sample. The third point is that low high school graduation rates are not the only reason why Hispanics are underrepresented in the military today. Hispanics are underrepresented among military accessions even when we restrict our sample to high school graduates in the Census. This is especially true when we consider the fact that Hispanics come disproportionately from poorer families, and more youth from poorer families enlist in the military than do those from wealthier families.

Hispanic High School Graduation Rates. Hispanics in the overall youth population graduate from high school at substantially lower rates than do either non-Hispanic whites or blacks. This is evident in Table 3.2, which shows the percentage of youth ages 17–21, not currently attending high school, who have earned either a high school diploma or GED. According to the Census, only 51 percent of Hispanic males and 64 percent of Hispanic females ages 17–21 have obtained a high school diploma. This percentage is sharply lower than that for either whites or blacks. Given that the services require enlistees to have earned a high school diploma, this gap in educational attainment goes a long way toward explaining why Hispanics enlist at much lower rates than either whites or blacks.

Table 3.2
Percentage of Youth Ages 17–21 with a High School Diploma,
by Race/Ethnicity and Gender

Race/Ethnicity	Census		NLSY97	
	Male	Female	Male	Female
White	84	88	85	87
Black	68	78	70	79
Hispanic	51	64	74	80

SOURCES: 2000 Census; NLSY97.

NOTES: Sample restricted to those not currently attending high school.
Cells report weighted means.

In the NLSY97, however, Hispanics complete high school at a much higher rate; 77 percent of Hispanic males and 82 percent of Hispanic females ages 17–21 surveyed in the NLSY97 have obtained a high school diploma. These graduation rates are equal to or higher than graduation rates for blacks, but still considerably lower than graduation rates for whites. Note that graduation rates for whites and blacks are very similar in the Census and the NLSY97, which suggests the Hispanic sample in the NLSY97 is particularly select.

Table 3.3 provides suggestive evidence that the difference in Hispanic high school graduation rates between the two surveys relates to immigration status. In Panel B of Table 3.3, we see that 45 percent of Hispanics ages 17–21 in the Census were born abroad, compared with 24 percent of Hispanics in the NLSY97. Among those born abroad, 84 percent of Census Hispanics are non–U.S. citizens (Panel C) and 77 percent immigrated to the United States within ten years of the survey (Panel D). In the NLSY97, these percentages are much lower (62 and 24 percent, respectively). So, not only are Hispanics surveyed in the Census much less likely to be U.S. born, those who are not U.S. born are much more likely to have recently immigrated to the United States. The NLSY97 does not deliberately exclude recent Hispanic immigrants from its sample and follows sampling procedures similar to that used by the Census. A longitudinal survey of this nature, though, is likely to garner poor response rates from this population. The survey itself demands a considerable amount of time to complete and is perhaps more invasive than recent Hispanic immigrants will tolerate. English proficiency is not a requirement for participating in the NLSY97, but lack of proficiency might cause many recent Hispanic immigrants to refuse participation during initial screening. The Census, on the other hand, is delivered by mail, asks a much more limited set of questions, and requires a short amount of time to complete. Census enumerators may be more successful in securing survey responses for those who fail to respond to the initial mail survey.

Observable differences in immigration status between the Census and NLSY97 Hispanic samples can account for only part of the differences we observe in high school graduation rates. This is evident if we examine high school graduation rates of recent immigrants in Table 3.3. Only 41 percent of non-U.S. born Hispanics surveyed by the Census had completed high school at ages 17–21, compared with 72 percent of non-U.S.-born Hispanics interviewed in the NLSY97 (Panel B). The disparity in high school graduation rates between the Census and NLSY97 also persists when we restrict our sample to non–U.S. citizens, Hispanics who immigrated within the last ten years, and Hispanics who speak Spanish at home.

Table 3.3
Immigration Status and High School Graduation Rates in the 2000
Census and NLSY97 (%)

	Race/Ethnicity		
	White, Non-Hispanic	Black, Non-Hispanic	Hispanic
A. High school diploma			
Census	86	73	57
NLSY97	86	75	77
B. Non–U.S. born			
Total			
Census	3	7	45
NLSY97	2	3	24
High school diploma			
Census	90	84	41
NLSY97	88	88	72
C. Non–U.S. citizen[a]			
Total			
Census	54	61	84
NLSY97	42	28	62
High school diploma			
Census	88	82	37
NLSY97	88	97	70
D. Immigrated within last 10 years[a]			
Total			
Census	58	65	77
NLSY97	45	36	24
High school diploma			
Census	88	83	35
NLSY97	88	67	58
E. Spanish spoken at home			
Total			
Census	3	3	76
NLSY97	1	1	69
High school diploma			
Census	88	84	53
NLSY97	92	90	72

SOURCES: 2000 Census; NLSY97.

NOTES: Sample restricted to individuals ages 17–21 not currently attending high school. Cells report weighted means.

[a] Sample further restricted to non–U.S. born.

Conversely, restricting our sample to U.S.-born Hispanics, we find that high school graduation rates computed in the Census and NLSY97 differ comparatively little (70 percent versus 78 percent) (not shown). Thus, even though recent immigrants make up a much larger percentage of Hispanics surveyed in the Census than surveyed in the NLSY97, that difference in sample composition can explain only a small percentage of the difference we observe in high school graduation rates because recent Hispanic immigrants in the NLSY97 have only slightly lower graduation rates than other Hispanics in the NLSY97.

We conclude from Table 3.3 that recent Hispanic immigrants in the Census and NLSY97 must differ along other dimensions that we cannot observe in one or both surveys. Differences in country of origin, which are not well measured in the NLSY97, are one possibility. It is also possible that recent Hispanic immigrants willing to participate in the NLSY97 come, on average, from better-educated and higher-income families than do other recent Hispanic immigrants. Willingness to participate in the NLSY97 may be correlated with certain personality characteristics and contextual factors that also correlate with educational attainment, such as willingness to trust survey personnel or living in a safe neighborhood.

Whatever the reason, it is clear that NLSY97 Hispanics are different from Census Hispanics with respect to educational attainment. We also note that these same differences exist between Hispanics interviewed in the 2003 Youth Poll and Census. The analyses in the subsections on weight, number of dependents, and substance abuse and illegal activity are largely limited to high school graduates. This sample restriction allows us to generalize from the NLSY97 and 2003 Youth Poll samples to the national population of Hispanic high school graduates. We caution, however, that the NLSY97 sample of Hispanic high school graduates may differ from the U.S. population in ways that are correlated with the enlistment standards we examine.

High School Graduation and Military Service. The final point we make in this subsection is that differences in high school graduation rates explain much, but not all, of the difference in military accessions we observe across racial and ethnic backgrounds. In Table 3.4, we show the percentage of youth ages 17–21 who have ever served in the U.S. military by race/ethnicity, gender, and educational attainment. Overall, 2.1 percent of Hispanic males ages 17–21 report having ever served on active duty, compared with 3.1 percent of white males and 5.1 percent of black males. When we condition the sample on high school graduates, Hispanics are almost as likely as whites to have ever served (3.0 percent versus 3.3 percent). Black high school graduates are considerably more likely to have served than either whites or Hispanics (5.8 percent). Thus, while high school graduation appears to close the gap in military service between whites and Hispanics, it has no effect on the gap in military service between blacks and Hispanics. This is true for both males and females.

There are many reasons why Hispanic high school graduates might enlist at lower rates than black high school graduates do. However, it is not immediately obvious from survey data that Hispanics have a lower desire to enlist in the military service than do blacks. In Panel B of Table 3.4, we tabulate the percentage of NLSY97 respondents who report they are likely or very likely to enlist in the military. As can be seen, blacks and Hispanics are equally likely to report that they expect to enlist in the military. While much can happen between the times expectations are reported and youth make definite decisions regarding military service, it is apparent that Hispanics, at least initially, are as favorably disposed to the possibility of military service as are blacks.

The discrepancy between expectations and realizations raises the question of why Hispanics are less likely to realize their military expectations than are blacks. In the following subsections, we explore whether enlistment standards have any explanatory power in this regard.

Table 3.4
Percentage in Military Service, by Race/Ethnicity,
Educational Attainment, and Gender

Race/Ethnicity	All	High School Diploma
A. Military service		
Males		
White	3.1	3.3
Black	5.1	5.8
Hispanic	2.1	3.0
Females		
White	1.3	1.3
Black	2.9	3.0
Hispanic	1.1	1.3
B. Enlistment expectations[a]		
White	13	2
Black	17	15
Hispanic	17	16

SOURCES: 2000 Census; NLSY97.

NOTES: Sample in Panel A is restricted to individuals ages 17–21 not currently attending high school. Panel B is restricted to individuals ages 15–19. Cells report weighted means.

[a] Percentage who report that they are likely or very likely to enlist in the military.

AFQT

We begin our investigation of the effect of enlistment standards on Hispanic eligibility by examining AFQT scores. Obtaining a minimum AFQT score is among the "hardest" constraints faced by potential enlistees. DoD derives the AFQT score from a subset of the Armed Services Vocational Aptitude Battery (ASVAB) that tests word knowledge, paragraph comprehension, math knowledge, and arithmetic reasoning. The ASVAB is administered to all potential recruits and is used to make occupational assignments as well as to determine eligibility for enlistment. DoD categorizes potential recruits according to eight AFQT categories. Categories V, IVC, IVB, IVA, IIIB, IIIA, II, and I corresponding to the 9th, 15th, 20th, 30th, 49th, 64th, and 92nd percentiles of a nationally normed AFQT distribution.

By law, no service may accept a Category V recruit, and only 4 percent of all recruits in a given fiscal year can be Category IV. In practice, the services accept very few Category IV recruits. In FY03, fewer than one half of one percent of Army and Marine Corps accessions were Category IV. The Air Force and Navy enlisted virtually no Category IV individuals.

In addition, the individual services maintain their own minimum AFQT requirements. For traditional high school graduates, minimum AFQT scores (equal to percentiles) are 16, 40, 31, and 21 for the Army, Air Force, Navy, and Marine Corps, respectively. Individuals holding a nontraditional high school degree (including GED holders) face stricter AFQT requirements.[5] The AFQT has recently been renormed, but, at the time this research was conducted, the new norms had not yet been applied to raw test score data available in the NLSY97. Consequently, we devised a method for norming AFQT scores in the NSLY97 that corresponds closely to the approach used by DoD. First, we obtained the residuals from a regression of each of the four ASVAB component scores on age in months and age in months squared. We then perform a principal components analysis on the four age-adjusted ASVAB components. Principal components analysis transforms a number of typically correlated variables into a smaller number of uncorrelated variables, called principal components, while maintaining the variability in the original variables. More specifically, a principal component is a linear combination of the original standardized ($\mu = 0$, $\sigma = 1$) variables (in this case, the four age-adjusted ASVAB components listed above) that maximizes the variance of that combination and has a zero covariance with all prior principal components. We take the first principal component as our measure of AFQT. We then weight the first principal component using the NLSY97 sampling weights to arrive at a nationally normed AFQT index.[6] Using this index, we then determine the percentile AFQT score for each respondent in our sample.

On average, blacks and Hispanics score lower on the AFQT than do whites. This is evident in Figure 3.1, which shows the cumulative percentage of our sample scoring below a given AFQT percentile by race/ethnicity. In our sample, 74 percent of blacks and 69 percent of Hispanics score below the 50th percentile on the AFQT compared with 40 percent of whites. Figure 3.2 shows the percentage scoring at or above the AFQT percentiles corresponding to DoD's classification scheme. Whereas 80 percent of whites would be classified at Category IIIB or above, only 49 percent of blacks and 53 percent of Hispanics would achieve Category IIIB or above.

AFQT scores of blacks and Hispanics have improved considerably relative to AFQT scores of whites over time. In our sample from the NLSY79 cohort an even larger percentage of blacks and Hispanics scored below the 50th percentile on the AFQT: 90 percent of blacks and 82 percent of Hispanics scored below the 50th percentile on the AFQT compared with 48 percent of whites. Figure 3.3 shows the percentage scoring at or above the AFQT percentiles corresponding to DoD's classification scheme in the NLSY79 cohort.

Conditioning our sample on high school graduates (by which we mean individuals who hold a traditional high school degree, GED, or alternative high school credential) lessens the gap in AFQT scores between whites and blacks and Hispanics, but whites continue to score considerably higher. In Figure 3.4, we see that the percentages of black and Hispanic high school graduates scoring below the 50th percentile fall to 62 and 66; 34 percent of white high school graduates score below the 50th percentile. In terms of AFQT categories, 85 percent of white high school graduates score in Category IIIB or above, compared with 6 percent of blacks and 62 percent of Hispanics (Figure 3.5). Figure 3.6 shows the percentage of high school graduates scoring in each AFQT category by race/ethnicity.

---

[5]   Holders of nontraditional high school diplomas must score above 30 in the Army and Marine Corps and above 50 in the Air Force and Navy.

[6]   Our AFQT index is based on the NLSY97 cross-sectional sample.

**Figure 3.1
Cumulative Density of AFQT Scores by Race/Ethnicity**

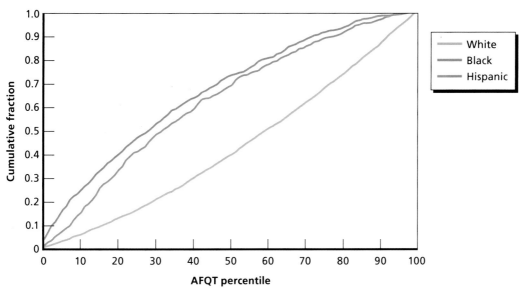

SOURCE: NLSY97.
NOTE: Sample restricted to individuals ages 17–21 not currently enrolled in high school.
RAND *MG773-3.1*

**Figure 3.2
Percentage Scoring At or Above Given AFQT Category, by Race**

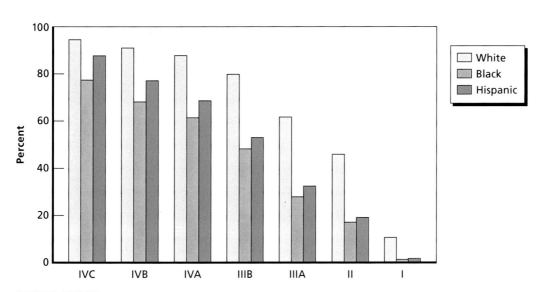

SOURCE: NLSY97.
NOTE: Sample restricted to individuals ages 17–21 not currently enrolled in high school.
RAND *MG773-3.2*

**Figure 3.3
Percentage Scoring At or Above Given AFQT Category, by Race: NLSY79 Cohort**

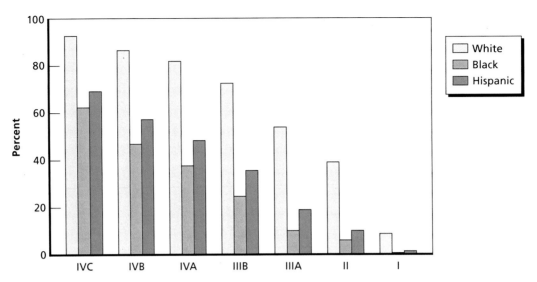

SOURCE: NLSY79.
NOTE: Sample restricted to individuals who are ages 17–21 not currently enrolled in high school.
RAND *MG773-3.3*

**Figure 3.4
Cumulative Density of AFQT Scores, by Race/Ethnicity: High School Graduates**

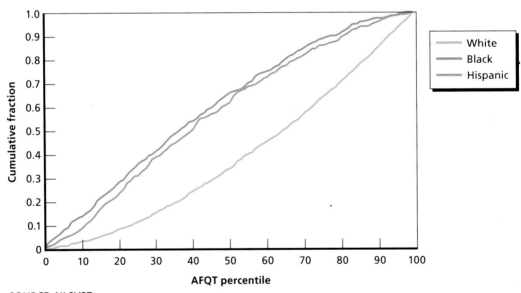

SOURCE: NLSY97.
NOTE: Sample restricted to individuals ages 17–21 who have a high school diploma.
RAND *MG773-3.4*

Figure 3.5
Percentage Scoring At or Above Given AFQT Category, by Race/Ethnicity: High School Graduates

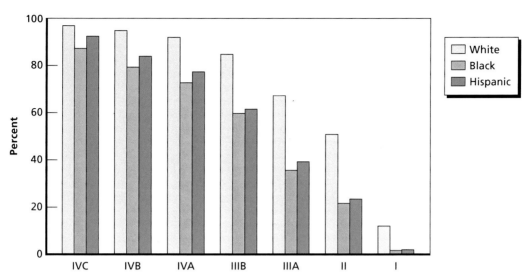

SOURCE: NLSY97.
NOTE: Sample restricted to individuals ages 17–21 who have a high school diploma.
RAND *MG773-3.5*

Figure 3.6
Percentage Scoring in a Given AFQT Category, by Race/Ethnicity: High School Graduates

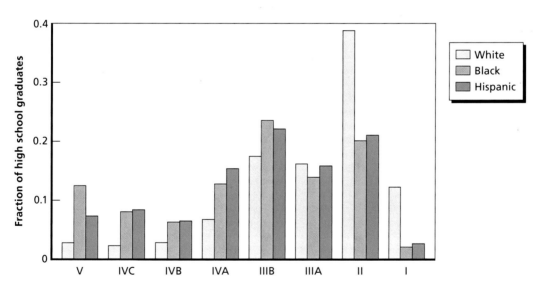

SOURCE: NLSY97.
NOTE: Sample restricted to individuals ages 17–21 who have a high school diploma.
RAND *MG773-3.6*

In the sections to follow, we condition our sample on high school graduates. It is worth considering, however, whether this conditioning is more or less restrictive than conditioning on the service's minimum AFQT scores. Figure 3.7 suggests that AFQT may be the more binding constraint for potential black and Hispanic recruits. Among Hispanics ages 17–21

Figure 3.7
Percentage with High School Diploma Meeting Minimum AFQT Requirements, by Race/Ethnicity and Service

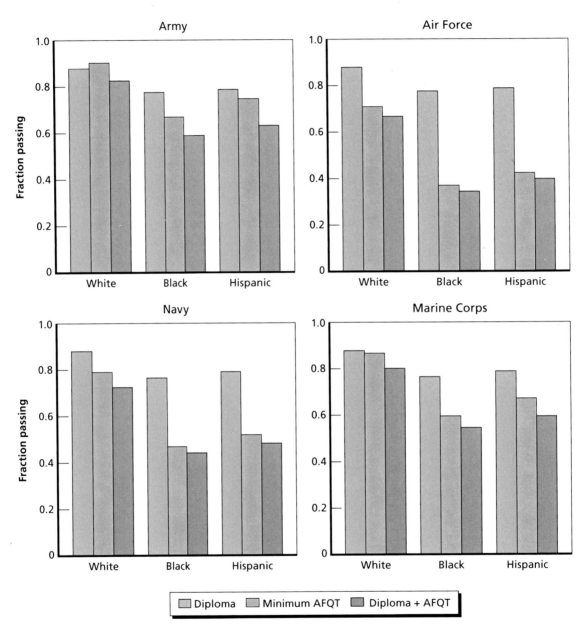

SOURCE: NLSY97.
NOTE: Sample restricted to individuals ages 17–21 who are not currently enrolled in high school.
RAND *MG773-3.7*

who are not currently attending school in our sample, 79 percent have a high school diploma.[7] Only 41 percent of Hispanics, however, achieve the Air Force's minimum AFQT score of 40.[8] Consequently, conditioning on high school graduation subsequent to conditioning on AFQT has little impact on the percent of eligible Hispanic recruits. In other words, a large percentage of Hispanic high school graduates (59 percent) fail to meet the Air Force's minimum AFQT requirement. This is also true in the Navy, whose minimum AFQT score is 31. In the Army and Marine Corps, minimum AFQT scores are less binding, but still represent a more significant obstacle to enlistment than does high school graduation.

One implication of Figure 3.7 is that increasing high school graduation rates among Hispanics may not lead to comparable increases in the enlistment-eligible Hispanic population, because high school graduation itself may not lead to a sufficiently large increase in AFQT scores in this population. For example, if 39 percent of Hispanic high school graduates in the NLSY97 currently fail to meet the Navy's minimum AFQT score of 31, we should expect no more than this percentage, and quite likely less than this percentage, to pass this minimum AFQT score were high school graduation rates to rise in the Hispanic population generally.

We make two more observations here before continuing on to investigate the effect of enlistment standards other than high school graduation and AFQT. First, while all but the Air Force's minimum AFQT requirements allow for Category IV enlistments, the services rarely in fact recruit such individuals. Thus, for all the services, the de facto AFQT requirement is to score in Category IIIB and above. Only 60 percent and 62 percent of black and Hispanic high school graduates in our sample, respectively, are Category IIIB and above. Also, the military occupations available to potential recruits depend on AFQT; individuals who score poorly on the AFQT might not like the occupations available to them. Blacks and Hispanics might dropout of the enlistment process at higher rates not only because a higher percentage fail to meet the minimum requirement, but because a much higher percentage of these individuals fail to achieve desirable AFQT scores from the perspective of the services and the recruits themselves.

The second observation is that the Hispanic sample in the NLSY97 might not be representative of the overall Hispanic population, as suggested by the comparisons between the Census and NLSY97 earlier in this chapter. According to the NLSY97, Hispanics graduate from high school at much higher rates than the Census shows; whatever selection drives this discrepancy might also cause NLSY97 Hispanics, even high school graduates, to score higher on the AFQT than do Hispanics in the general population. Thus, while black and Hispanic high school graduates in the NLSY97 appear to have comparable AFQT scores, it may be that Hispanic high school graduates in the general population score lower on the AFQT than is indicated here.

---

[7] Table 3.3 reports that 77 percent of Hispanics ages 17–21 not in high school have a high school diploma. This figure differs from the 79 percent stated in the text because the numbers in Table 3.3 include individuals who have missing enlistment criteria. Figure 3.7 excludes those individuals.

[8] As noted above, minimum AFQT scores depend on the type of high school credential held. The score used in the second column of these figures (Minimum AFQT) is the score applicable to traditional high school graduates. The third column (Diploma+AFQT) applies the minimum AFQT score for the particular type of diploma held by a given individual.

Weight

Weight and number of dependents are, by themselves, significant obstacles to enlistment for a high percentage of youth. Weight standards, especially, are growing in importance because of the startling growth over the past 20 years in the percentage of youth classified as overweight or obese (Ogden et al., 2002). The number of dependents is less of a factor, but it is nonetheless important for Hispanics, who tend to have more children than non-Hispanics and have them at an earlier age. For example, in the NLSY79, 43 percent of Hispanic women have at least one child by age 21 compared with 24 percent of white women and 51 percent of black women. By age 40, NLSY79 Hispanic women have an average of 2.5 children, compared with 1.8 and 2.2 children for white and black women.

Each of the services maintains its own enlistment standards with respect to weight and family size. Weight standards are defined in terms of weight-for-height (there is also a minimum height standard). The Marine Corps employs the most lenient weight standard but also requires its recruits to pass a strength test, which presumably accounts for muscle mass. The Army has the most stringent weight standard. The Marine Corps employs a more lenient weight standard when individuals first enlist and a more stringent standard when recruits exit basic training. The analyses below are based on enlistment rather than shipping weight standards.

Data from the NLSY indicate that American youth are significantly heavier today than they were a few decades earlier. We find that mean weight, conditional on height and age, increased by 12.5 and 10.3 pounds between 1980 and 2001 among male and female youth ages 17–21. As seen in Figure 3.8, which presents the cumulative distribution of weight (adjusted for height and age), the increase in weight occurred throughout the weight distribution for both males and females. This indicates that it is not just that the heaviest individuals are getting even heavier, but that youth of all weights are getting heavier.[9]

The NLSY data also indicate that, controlling for height, blacks and Hispanics are heavier than whites. Among youth ages 17–21 in 2001, black and Hispanic males weigh on average 6.3 and 9.7 pounds more than white males; black and Hispanic females weigh on average 16.7 and 7.6 pounds more than white females. Figure 3.9 shows that blacks and Hispanics are heavier than whites throughout the weight distribution.

Given the difference in weight we observe between whites, blacks, and Hispanics in Figure 3.9, it comes as no surprise that a higher percentage of blacks and Hispanics fail to meet the military's weight standards. This is apparent in Figures 3.10 and 3.11 which, in the first bar of each set, graph the percentage of male and female youth passing weight standards by service and race/ethnicity. Depending on the service, between 79 and 91 percent of white males pass weight standards, compared with between 69 and 86 percent of black males and between 71 and 88 percent of Hispanic males. A considerably smaller percentage of female youth passes military weight standards. Depending on the service, between 63 and 82 percent of white females pass weight standards, compared with between 42 and 65 percent of black females and between 49 and 71 percent of Hispanic females.

---

[9] Figure 3.8 was generated by taking the residual from a linear regression of weight on height, height squared, age, and age squared. The regressions were implemented separately by gender. The residuals were then graphed against percentiles by cohort (1979 and 1997).

Figure 3.8
Cumulative Distribution of Adjusted Weight, by Youth Cohort

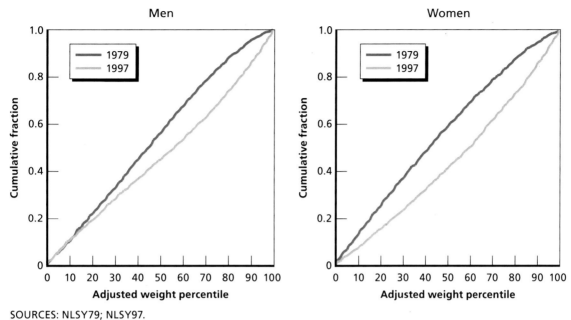

SOURCES: NLSY79; NLSY97.
NOTE: Sample restricted to individuals ages 17–21.
RAND *MG773-3.8*

Figure 3.9
Cumulative Distribution of Adjusted Weight, by Race/Ethnicity

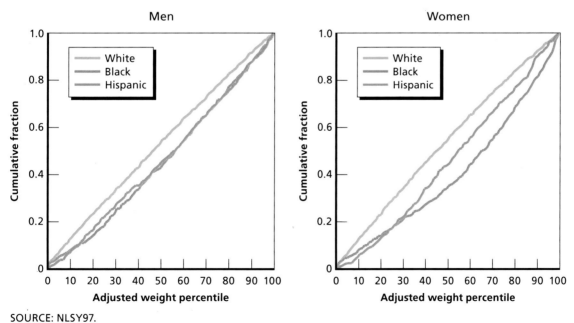

SOURCE: NLSY97.
NOTE: Sample restricted to individuals ages 17–21.
RAND *MG773-3.9*

Figure 3.10
Percentage of Males Meeting Weight, High School/AFQT, and Combined Enlistment Standards, by Service and Race/Ethnicity

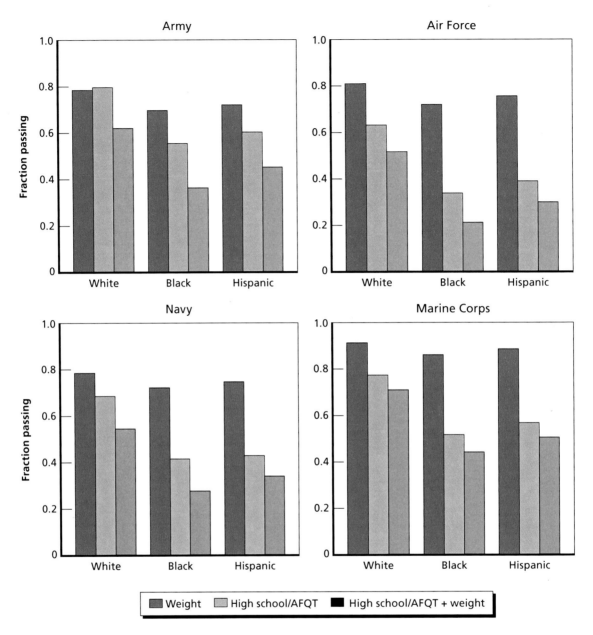

SOURCE: NLSY97.
NOTE: Sample restricted to individuals ages 17–21 who are not currently enrolled in high school.
RAND MG773-3.10

Figures 3.10 and 3.11 also make clear that weight has an independent effect on the cumulative percentage passing enlistment standards. The final bar of each set in each graph indicates that even among individuals with a high school diploma and who score above the minimum

Figure 3.11
Percentage of Females Meeting Weight, High School/AFQT, and Combined Enlistment Standards, by Service and Race/Ethnicity

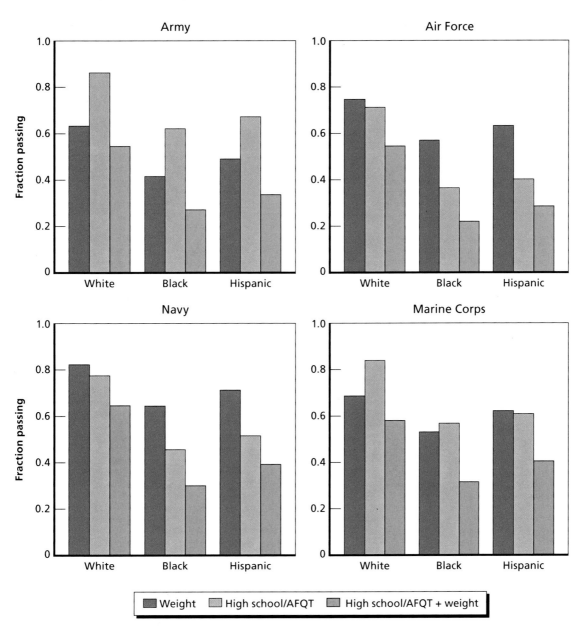

SOURCE: NLSY97.
NOTE: Sample restricted to individuals ages 17–21 who are not currently enrolled in high school.
RAND *MG773-3.11*

on the AFQT, a considerable percentage fail to meet weight standards. This appears to be equally true across race/ethnic groups. For example, about one-third of males meeting the minimum AFQT standard would fail to meet the Army's weight standard.

Number of Dependents

With respect to family size, the Army and Air Force prohibit enlistees from having more than three dependents, including a spouse. The Marine Corps and Navy prohibit enlistees from having more than one dependent. A biological child is considered to be a dependent regardless of whether the child lives with the parent, and no service allows a single custodial parent to enlist. The services differ in how they define dependency in cases in which individuals are separated from their spouse or in cases in which an individual has adopted children or stepchildren. For the purposes of this report, we define a dependent child as any biological or adopted child.

Blacks and Hispanics are more likely to have children at an early age and so are less likely to meet the military's enlistment standards with respect to dependents. In our sample of youth ages 17–21, 24 percent of blacks and 20 percent of Hispanics have children, compared with only 9 percent of whites. A much higher percentage of black youth with children are unmarried; 93 percent of black youth with children are unmarried, compared with 77 percent of Hispanic and 73 percent of white youth with children.

Overall, the impact of the dependents requirement is less significant than the impact of weight, especially among those who meet minimum AFQT standards. Among males, for example, between 94 and 96 percent of whites, between 85 and 86 percent of blacks, and 90 percent of Hispanics meet the dependents standards (see Figure 3.12, first bar in each set). Among females, we observe a higher percentage of youth failing to meet the dependents standards and greater differences across race/ethnic groups; between 87 and 90 percent of white females, between 71 and 73 percent of black females, and between 75 and 81 percent of Hispanic females fail to meet the dependents standards (see Figure 3.13). Reporting error most likely accounts for the difference in the percentage of male and female youth passing the dependents standards. Males are much less likely than females to report accurately the number of children they have.

Once we condition on meeting high school diploma and AFQT requirements, the impact of the dependents standards is even less. Fewer than 5 percent of male and 12 percent of female high school graduates who meet minimum AFQT standards fail to meet the dependents standards. This should come as no surprise, since educational attainment and early childbearing are highly correlated.

Substance Abuse and Illegal Activity

The armed forces require that persons entering military service be of good moral character. In an effort to meet this goal, each service maintains detailed enlistment standards with respect to substance abuse (past or current) and past illegal activity. Waivers are frequently granted to individuals who fail to meet these standards, although failure to meet some categories of substance abuse and illegal activity permanently disqualifies individuals from enlistment. For example, the services will not enlist any individual who fails a drug and alcohol test at the time of enlistment. Any applicants with court charges filed or pending against them (including both criminal and juvenile), and any applicants under civil restraint (including confinement, parole, and probation) are disqualified and not eligible for waivers. Additionally, felons are categorically excluded from service.[10]

---

[10] However, the Secretary of a given service may authorize "exceptions in meritorious cases, for the enlistment of . . . persons convicted of felonies" (DoD Directive 1304.26, p. 8).

Figure 3.12
Percentage of Males Meeting Dependents, High School/AFQT, and Combined Enlistment Standards, by Service and Race/Ethnicity

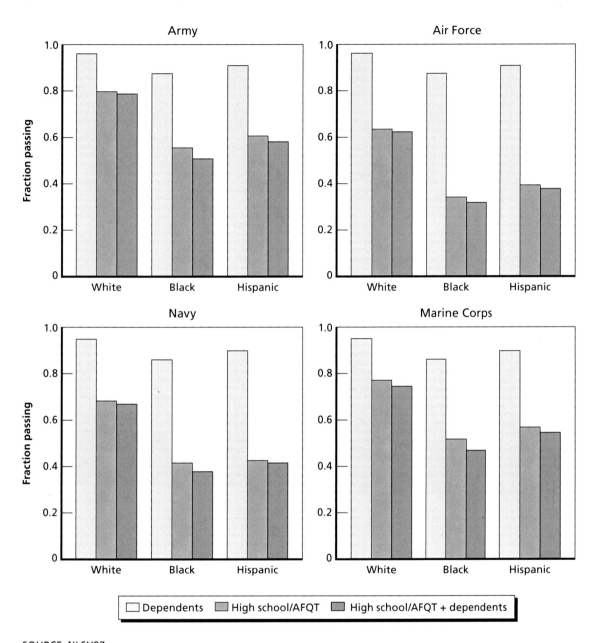

SOURCE: NLSY97.
NOTE: Sample restricted to individuals ages 17–21 who are not currently enrolled in high school.
RAND MG773-3.12

Potential enlistees are required to disclose past drug and alcohol use and illegal activity during the enlistment process. In addition, MEPS personnel conduct an Entrance National Agency Check (ENTNAC) on all applicants. This check includes offenses that have been expunged or cleared from a person's record. Each recruit is informed about the depth of the criminal record check used during the enlistment process and is required to disclose such inci-

Figure 3.13
Percentage of Females Meeting Dependents, High School/AFQT, and Combined Enlistment Standards, by Service and Race/Ethnicity

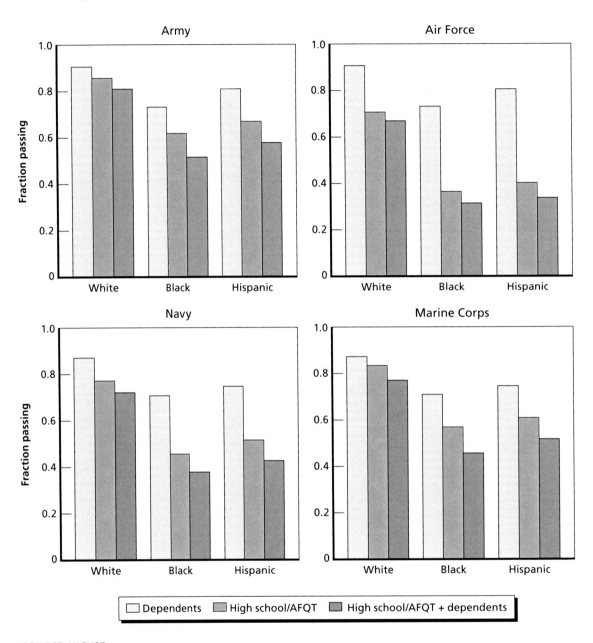

SOURCE: NLSY97.
NOTE: Sample restricted to individuals ages 17–21 who are not currently enrolled in high school.
RAND *MG773-3.13*

dents. A failure on the part of the applicant to disclose any information regarding criminal record may result in disqualification.

The NLSY queries respondents about drug and alcohol use and engagement in past illegal activities. There are three problems with these data. First, the data are self-reported, and, while respondents are told that their responses are confidential, they may not respond as truthfully to these questions as they would when interviewed at the MEPS. Thus, it is reasonable to assume that drug use and illegal activity are underreported in the NLSY. Second, the NLSY questions on these topics do not mesh particularly well with the services' particular enlistment standards. Finally, we know that waivers for past drug and alcohol use and misdemeanors are common; consequently, whether an individual has ever used drugs or committed a crime does not provide sufficient information to determine whether he or she would be rejected for enlistment.

In 2003, DoD's Joint Advertising Market Research and Studies (JAMRS) program fielded a youth poll that specifically addressed compliance with moral standards. The questions on drug use and criminal activity fielded in that survey are more closely aligned with actual enlistment standards. In Figure 3.14, we graph the percentage of youth ages 17–21 not currently attending high school who answered negatively to the following questions: (1) Have you ever been convicted of a felony? (2) Are you currently under any form of judicial restraint such a bond, awaiting trial, probation, or parole? (3) Have you ever been convicted of a misdemeanor? (4) Are you or have you ever been dependent on drugs or alcohol? We also graph in Figure 3.14 the percentage who respond affirmatively to the question: If you took a drug test, do you think you would pass today?

The raw statistics indicate that 90 percent of males and 98 percent of females overall are not under a judicial restraint and have never been convicted of felony. A higher percentage of youth—about 20 percent of males and 7 percent of females—have been convicted of a misdemeanor. About 83 percent of males and 89 percent of females reported that they would pass a drug test if given at the time of the interview. There appears to be little variation in self-reported drug use and illegal activity across race/ethnic groups.

The right-hand panels in Figure 3.14 indicate that judicial restraints and felonies have a minor impact on enlistment eligibility once we condition the sample on high school graduation (but not minimum AFQT, which is not available in the JAMRS youth poll). However, a substantial percentage of male high school graduates (19 percent) report that they have been convicted of a misdemeanor or would not pass a drug test if given today (16 percent). As emphasized above, whether these last two restrictions would significantly limit the enlistment pool is unclear, since misdemeanors are frequently waived and many individuals would presumably curtail drug use if they knew they would have to take a drug test.

The Cumulative Effect of Enlistment Standards

Altogether, a relatively small percentage of youth in our sample is eligible to enlist in the military. This is seen in Figures 3.15 and 3.16, which show the cumulative impact of enlistment standards on the percentage eligible to enlist by race/ethnicity, gender, and service and accounts for the percent of NLSY97 respondents reporting they have used an illegal drug in the past year or ever been convicted of a crime. By these estimates, only 45, 32, and 35 percent of white, black, and Hispanic males are eligible to enlist in the Marine Corps, the service with the cumulatively least stringent enlistment standards. Even ignoring moral requirements, only 68, 41, and 48 percent of white, black, and Hispanic males are eligible to enlist in the Marine Corps. The corresponding percentages for white, black, and Hispanic females are 35, 22, and 24 including moral standards, and 55, 26, and 36, not including moral standards.

Figure 3.14
Percentage Reporting No Illegal Activity or Substance Abuse, by Race/Ethnicity

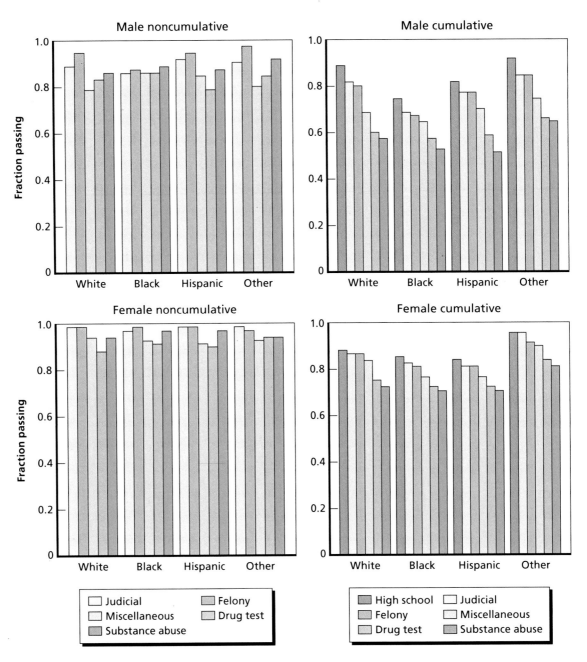

SOURCE: 2003 Youth Poll.
NOTE: Sample restricted to individuals ages 17–21 who are not currently enrolled in high school.
RAND *MG773-3.14*

Figure 3.15
Cumulative Impact of Enlistment Standards on Percentage Eligible to Enlist, by Race/Ethnicity: Males

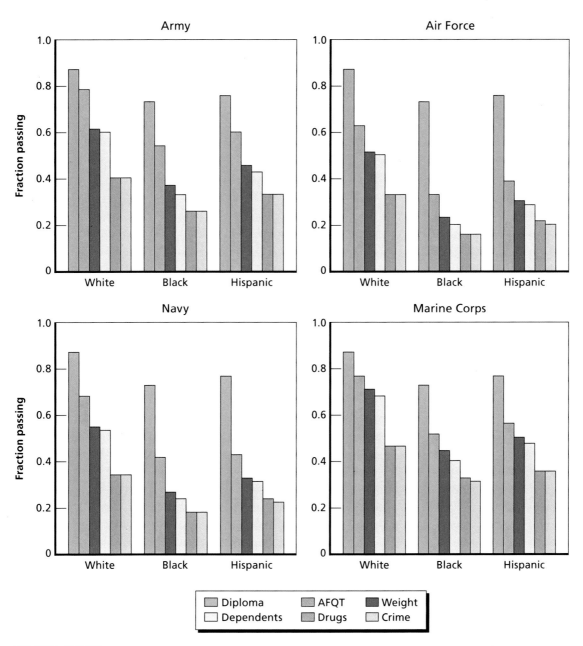

SOURCE: NLSY97.
NOTE: Sample restricted to individuals ages 17–21 who are not currently enrolled in high school.
RAND *MG773-3.15*

**Figure 3.16
Cumulative Impact of Enlistment Standards on Percentage Eligible to Enlist, by Race/Ethnicity:
Females**

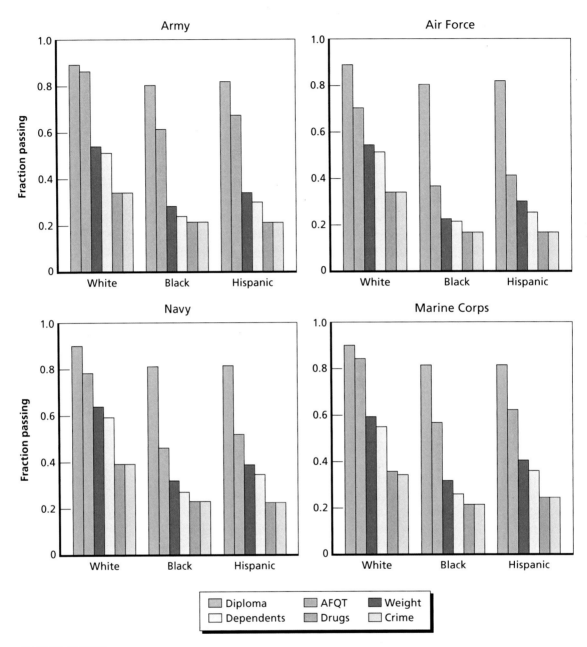

SOURCE: NLSY97.
NOTE: Sample restricted to individuals ages 17–21 who are not currently enrolled in high school.
RAND *MG773-3.16*

One might naturally ask at this point, "Which enlistment standard limits eligibility the most?" Or, from a policy perspective, one might ask how relaxing a particular enlistment standard—say, by defining job-specific weight standards or accepting more Category IV recruits—would affect overall eligibility. Because eligibility is correlated across different standards, we know that relaxing one standard will not necessarily result in a proportional increase in the percentage of youth eligible to enlist. For example, if we relax AFQT standards, only a percentage of the individuals now eligible to enlist on the basis of AFQT will be eligible to enlist on the basis of all other enlistment criteria (e.g., high school graduation, weight). Put another way, individuals with relatively low AFQT scores are less likely to pass other enlistment standards than are individuals with relatively high AFQT scores.

In Figures 3.17–3.19, we show how changing AFQT and weight standards relative to each service's current standards would affect eligibility based on all eligibility standards combined. For these figures, we employ the services' de facto AFQT standards (31 and above in the Army, Navy, and Marine Corps, and 40 and above in the Air Force). In the case of the Army, Figure 3.17 shows that about 22 percent of Hispanic males would be eligible to enlist at current standards (indicated as "0" on the x-axis). If the Army relaxed its weight standard so that 10 percent more Hispanic males were now eligible to enlist based on the weight standard alone, the percentage eligible to enlist in the Army based on all standards combined would increase from the current 22 percent to only 25 percent. Thus, the figure shows that relaxing the weight standard results in a much less than proportional increase in overall eligibility. This is because eligibility is strongly positively correlated across individual enlistment standards. For example, many Hispanics who exceed the current weight standard are not high school graduates.

Figure 3.17 also shows that relaxing Army AFQT standards in order to admit 10 percent more Hispanics based on that standard alone would result in an increase in overall eligibility of close to 10 percent. This suggests that for the Army, the AFQT standard is more of a limitation than the weight standard. Relaxing the AFQT standard would result in a far greater change in overall eligibility than a comparable relaxation in the weight standard. In the case of Hispanic males, this appears to be true across all the services. The marginal impact of relaxing AFQT standards is greater than the marginal impact of relaxing weight standards. This can be seen by noting that the slopes of the lines labeled AFQT in Figure 3.17 are generally steeper (and closer to the 45 degree lines) than the slopes of the lines labeled weight. Incidentally, this is not true of white males (Figure 3.18), for whom relaxing AFQT and weight standards appears to have an equivalent and less than proportional effect on overall enlistment.

Figure 3.19 shows comparable data for Hispanic females. For these youth, AFQT and weight standards appear to be equally binding. Relaxing AFQT and weight standards relative to current standards results in far less than proportional increase in eligibility overall. Comparable figures for white and black females and black males can be found in Appendix C of Asch et al. (2005).

Figure 3.17
Impact of Changing AFQT or Weight Standard on Overall Eligibility: Hispanic Males

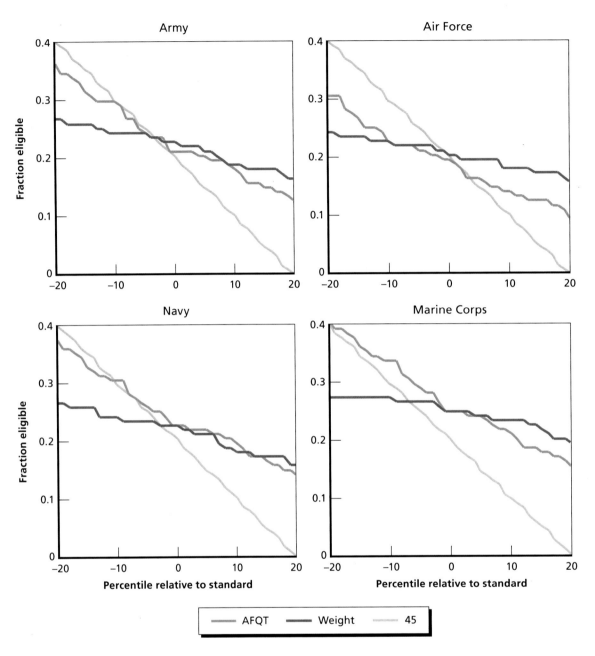

SOURCE: 2003 Youth Poll.
NOTE: Sample restricted to individuals ages 17–21 who are not currently enrolled in high school.
RAND *MG773-3.17*

Figure 3.18
Impact of Changing AFQT or Weight Standard on Overall Eligibility: White Males

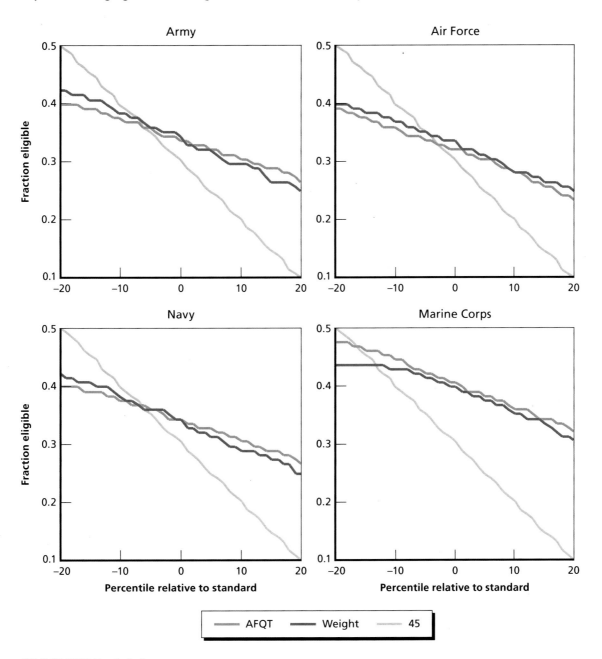

SOURCE: 2003 Youth Poll.
NOTE: Sample restricted to individuals ages 17–21 who are not currently enrolled in high school.
RAND *MG773-3.18*

Figure 3.19
Impact of Changing AFQT or Weight Standard on Overall Eligibility: Hispanic Females

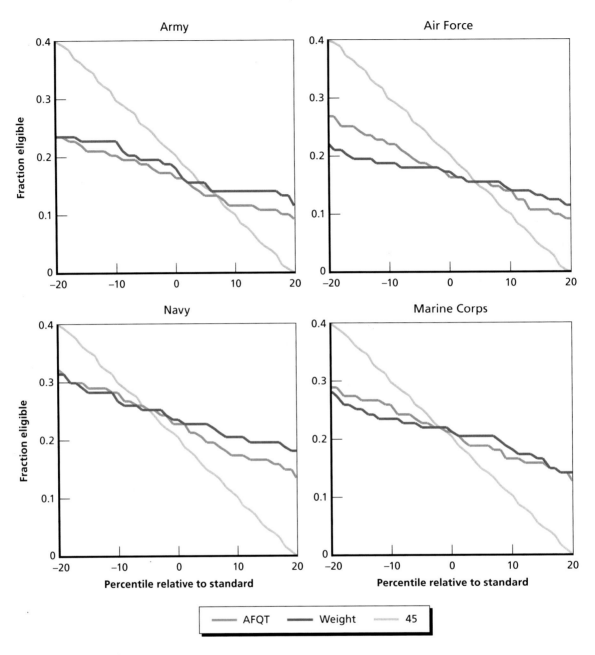

SOURCE: 2003 Youth Poll.
NOTE: Sample restricted to individuals ages 17–21 who are not currently enrolled in high school.
RAND *MG773-3.19*

Conclusions

Our analysis of the degree to which the youth population meets enlistment standards indicates that while failure to obtain a high school diploma is an important reason why Hispanics are underrepresented in the military, difficulty in meeting AFQT and weight standards is also an important factor limiting their eligibility. About 57 and 67 percent of male and female Hispanic high school graduates, respectively, fail to score above Category IV on the AFQT and meet the Army's weight standard. Thus, it is unlikely that increases in the rate of high school graduation among Hispanics alone will yield a military force that is fully representative of the Hispanic population. Our analysis above, in the subsection on the cumulative effect of enlistment standards, suggests that relaxing AFQT standards for Hispanic males (i.e., admitting more Category IV youth) would have the highest marginal impact on enlistment eligibility in that population. For Hispanic females, relaxing AFQT and weight standards would have equally strong marginal impacts on overall eligibility, reflecting the relative importance of weight standards in that population. These findings also generally apply to the black youth population.

Enlistment standards related to dependents appear to be relatively unimportant on their own. Very few Hispanic youth (male or female) who secure a high school diploma and meet minimum AFQT requirements are disqualified on the basis of marital status and family size. This research suggests that criminal activity and drug use could be important, but no more so for Hispanics than for whites or blacks. The impact of waivers, which we could not account for in our analysis, is likely to diminish the overall importance of criminal activity and drug use in limiting Hispanic enlistments.

The policy implications of our analyses are discussed in Chapter Seven.

CHAPTER FOUR

Health Obstacles to the Enlistment of Hispanic Youth

Health is a major criterion for both enlisting in the military and continuing to serve. As discussed in Chapter Two and Appendixes A and B, the screening process considers a large number of health factors. In this chapter, we discuss why we might expect health factors to screen out Hispanics disproportionately. The analysis supplements the analysis in Chapter Three. The purpose of health standards is to ensure that service men and women are medically able to carry out their responsibilities, not likely to be absent from duty because of a serious health problem, and not likely to compromise the health of others. Health is carefully screened before enlistment (and throughout service), and is a significant filter for many who desire to enlist.

Because the incidence of any given health condition is quite small in most cases, it is infeasible to assess the percentage of the population that is ineligible due to a specific health factor. Instead, the analysis groups health factors into three key aspects that are considered for enlistment: weight, major medical conditions, and minor medical conditions. The latter two grouping are based on whether an individual can receive a waiver.

Weight is screened using height-weight charts. However, height-weight charts are not the only criterion: when an individual fails to "make weight," other measures (such as waist, neck, and wrist measurements) may be used to help identify those who are heavy due to muscle mass rather than fat. Each service has its own weight standards; the Army is the most restrictive; the Air Force is the least restrictive.

Major health conditions are those that cannot be waived. Such conditions include, for example, having a functional limitation or a noncorrectable vision or hearing problem. Asthma (having an asthma attack within the past year) is considered a major condition. Because asthma in the Hispanic population is a serious concern for recruiters, we present the results for this condition separately.

Minor health conditions, at the discretion of the service, may be waived. Such conditions include, for example, skin allergies, hay fever, bone or joint surgery, and attention deficit disorder.

Previous Research on Hispanic Health

There is reason to suspect Hispanics may suffer a greater burden from many of the health conditions considered for enlistment. Hispanics are on average poorer and less educated than whites, and less educated on average than blacks (Hoffman, Llagas, and Snyder, 2003; Llagas, 2003). Because income and education are positively correlated with health, we would expect

Hispanic health patterns to reflect their relative socioeconomic patterns; in other words, we would expect Hispanics to have poorer health than whites or blacks.

However, previous research on Hispanic health has found mixed results, depending largely on the population chosen for analysis or the specific health condition measured. Hispanics do suffer disproportionately from obesity. Numerous studies have documented higher levels of overweight status and obesity among Hispanics, even among military personnel (Freid et al., 2003; Lindquist and Bray, 2001; Nolte et al., 2002; Popkin and Udry, 1998). Hispanics also suffer disproportionately from diabetes. However, Hispanics do not appear to be at a disadvantage regarding cancer, low birth weight, infant mortality, and a number of other important health outcomes (see Vega and Amaro, 1994, for a thorough review).

Researchers have identified a paradox of better-than-expected health outcomes in the Hispanic population, often referred to as the "Hispanic paradox." While the cause of the paradox is still being debated, many researchers have concluded that the large percentage of immigrants in the Hispanic population is the primary explanation for the Hispanic paradox (now often referred to as the "immigrant paradox").[1] In fact, recent studies have found a similar paradox for most ethnic groups when immigrant and nonimmigrant health outcomes are compared (Cho and Hummer, 2001; Franzini, Ribble, and Keddie, 2001; Frisbie, Cho, and Hummer, 2001; Palloni and Morenoff, 2001). Weight problems and diabetes are the notable exceptions to the immigrant paradox for Hispanics. However, poorer results for weight may be balanced by better outcomes in major and minor medical conditions. This portion of our analyses was designed to reveal whether the services' health criteria disproportionately screen out Hispanic youth, and, if so, the basis for the disqualification.

Citizenship and National Orgin

When referring to categories identifying race or ethnicity, it is important to consider that the term "Hispanic" is used to refer to a diverse group of people. In this chapter, we look specifically at two factors: citizenship (the services prefer recruits who are citizens; the career and reenlistment opportunities for noncitizens can be limited) and national origin.

Approximately 4 percent of enlisted members identify their ethnicity as "Mexican," the largest ethnic group among those reported by DoD (Department of Defense, 2006, Table B-25). Therefore, understanding the similarities and differences between this group and the larger all-Hispanic group is clearly important. Where possible, we have separately identified results for "Mexican" and "all-Hispanic."

---

[1]   The two main explanations for the paradox center on the selectivity of immigration and on the effects of acculturation. The immigration hypothesis suggests that only the healthiest individuals can undertake moving to another country, legally or especially illegally, resulting in better health among immigrants. The acculturation hypothesis suggests that immigrants start with healthier lifestyles, eschewing smoking, alcohol, and unhealthy "Americanized" diets. However, the longer immigrants live in the U.S., the more acculturated they become, taking on more unhealthy behaviors. The two hypotheses are not mutually exclusive, and there is evidence to support both. Research results are often highly dependent upon the ages considered and the specific health measure being used. See, in particular, Franzini, Ribble, and Keddie (2001) and Palloni and Morenoff (2001) for reviews of both perspectives.

Data

The data used for analysis are from the National Health Interview Survey (NHIS) from 1998 to 2001. The NHIS is a cross-sectional, household-based survey of diseases, health conditions, and behaviors in the civilian, noninstitutionalized population. The survey has been conducted annually since 1957 and serves as the principal source of information on the health of the U.S. population. The content of the survey is updated every 10–15 years.

The survey design oversamples blacks and Hispanics, ensuring enough cases to examine these minority populations. By pooling several years of data, we are able to examine national origin subpopulations with statistical precision.

Our analyses limit sampled data to those ages 18–25. This sample consists mainly of individuals usually considered to be in the prime recruiting ages. Our strategy for measuring enlistment standards is described in the following subsection.

Enlistment Standards and Measurement Strategy

Numerous health conditions may disqualify a person from military service. They range from clearly limiting major conditions, such as noncorrectable hearing or vision problems, to less severe minor conditions, such as skin allergies, hay fever or attention deficit disorder. The actual requirements for military service are quite complex and differ across the four branches (The specific requirements are summarized in Chapter Two).

Further, some disqualifying conditions are waiverable if the recruiter petitions for a waiver on behalf of the applicant. The waiver process depends on the decision of a commander at the local level or higher and on the nature of the waiver (i.e., waivers for more serious conditions must be granted by higher-echelon personnel). To some extent, the standards for granting a waiver are somewhat fluid and may depend on recruiting needs at the time. For this analysis, the set of service-specific major and minor health conditions has been thoroughly reviewed and matched as closely as possible with available measures in the NHIS data.

There is an inherent difficulty in matching the set of detailed military health requirements with information available in such large, representative datasets as the NHIS. The two do not mirror each other. For example, the list of health measures constructed from the NHIS does not encompass the entire set of military medical disqualifiers, nor does it perfectly measure the criteria applied by each individual branch of the military. Therefore, the measures used in this analysis are those that, on balance, most closely capture the set of disqualifying conditions in the military regulations (nonwaiverable or rarely waived, as well as waiverable), and represent the most serious or limiting, and the most prevalent, disqualifying health conditions. Consequently, these measures provide estimates of the percentages of the target populations that are *likely* to face obstacles to enlistment.

Measurement of Health Conditions

As mentioned above, we identified three general health measures, indicating whether an individual would likely be disqualified based on weight, a major health condition, or a minor condition. While not perfect, this measurement strategy captures most of those who would fail to meet the medical enlistment standards across the services. We aggregated individual conditions because of the relative infrequency of any single disqualifying health condition. There are simply not enough cases to examine each condition separately with any precision. A more detailed discussion of each measure follows. Prevalence rates for each of the specific conditions comprising the aggregate measures are reported in Appendix C.

Weight. We compared self-reported height and weight to the height-weight charts for each of the services, using new accession standards if they differed from active duty standards (only the Army and Marines have different standards for new accessions). In reality, however, when an applicant fails to "make weight," anthropometric measures (such as circumference of waist, neck or wrist) are used to identify individuals who may be heavy because of muscle mass rather than fat. The NHIS does not include anthropometric measures; consequently, we cannot identify those who might fail weight but pass the "tape test." Therefore, our weight measures may overstate the percentage of failures that are due to weight.

Major Health Conditions. These conditions are treated as nonwaiverable (according to published medical standards for each of the four branches). Major conditions are having a functional limitation; having a noncorrectable vision or hearing problem; needing special equipment for daily activity; having organ failure, heart condition, diabetes, ulcer, stroke, emphysema, or hypertension; or having an asthma attack in the past year. The last measure does not perfectly capture the military standard for asthma. For the military, active asthma after age 14 is considered nonwaiverable. But asthma before age 14 may be waiverable (for example, if the applicant passes a pulmonary function test). Our measures ask only two questions: "Have you ever been told by a doctor you have asthma?" and "During the PAST 12 MONTHS, have you had an episode of asthma or asthma attack?" We used the information about attacks in the past 12 months, treating this as an indicator of disqualification from enlistment, rather than using information about ever having been diagnosed with asthma. (Many individuals have been diagnosed with asthma as children but have not had active asthma since childhood.) Our approach potentially underestimates the number of individuals who might be disqualified because of asthma.

Minor Health Conditions. These conditions are listed as disqualifying but not specifically nonwaiverable. These conditions include sinusitis, hay fever, and chronic bronchitis (chronic respiratory conditions are an area of concern). Also included are other minor conditions, such as skin allergies, bone or joint surgery, headaches, and attention deficit disorder.

Approach

To understand the prevalence of disqualifying health conditions in the Hispanic population, we performed several analyses. First, using recent data, we measured the prevalence of a number of disqualifying conditions for non-Hispanic whites, non-Hispanic blacks, and Mexicans in the adult sample.[2] In addition, we also present the results for an aggregate "All-Hispanic" category.

The health condition questions in the NHIS are mostly self-reports of having ever been diagnosed by a doctor with a given condition. Consequently, this measure may not accurately capture the prevalence of a condition if there is underdiagnosis in certain populations for any reason. This may be especially relevant for the Hispanic population, which has much lower

---

[2]   Other Hispanic ethnic groups, such as Cubans or Puerto Ricans, do not have enough cases, even pooling multiple years of data, to give statistically precise estimates. The aggregate "All-Hispanic" category includes Mexicans, Puerto Ricans, Cubans, and other Hispanics. Therefore, the Hispanic category and Mexican category are not mutually exclusive.

rates of insurance and may face language and other barriers to obtaining high-quality medical care. As a consequence, individuals may have conditions that are undiagnosed either because they have not seen a doctor recently or because they have not received high-quality care. To address this concern, the results presented for health conditions are adjusted for having seen a doctor in the past 12 months. This approach is not ideal, however. Doctors' visits may be a poor proxy for access to health care, and there can be ethnic differences in doctors' tendency to diagnose some health problems.

We present results for males and females separately. Although the majority of new enlistees are male (roughly 85 percent across all services), it is also of interest to determine whether health serves as a substantial obstacle to the enlistment of women.

Results

This section present the results of our analyses. As discussed earlier, results are presented for several aggregate measures of enlistment-related health measures. The frequency of the specific health conditions that form the aggregate measures are presented in Appendix C.

In the following charts, each bar successively (i.e. cumulatively) restricts the sample considered. The first bar represents all those ages 18–25 (the military's prime recruiting population). The second bar represents only those 18-25-year-olds who are citizens. The third bar further limits the population to 18-25-year-old citizens who are also high school graduates. This same logic applies to all the charts presented for adults.

Because of the multiple comparisons that can be made, Appendix D presents the results of statistical significance tests of the difference across the restriction set, within ethnicity and across ethnicities, and within subpopulation. No adjustments were made for conducting multiple comparisons. Figure 4.1 shows that Mexican males are the most likely to be disqualified on the basis of weight. The pattern for all Hispanics roughly mirrors the results for Mexicans. Over 30 percent of young Mexican men fail to meet the weight standards. While white males are most likely to meet weight standards, the percentage of young men in each group who fail the weight standards is discouragingly high.

As we further restrict the population (first to those who are U.S. citizens, and then to those who are U.S. citizens with high school degrees), the prevalence of weight problems tends to rise. In other words, those individuals whom the military would prefer are less likely to pass the weight standards, regardless of ethnicity. Selection appears to act most substantially at the citizenship criterion for Mexicans and at the high school graduate criterion for blacks.

Figure 4.2 shows that among females, young black women are the most likely to fail the weight standards. Roughly 60 percent fail the weight standards of at least one branch. Considering that black women are overrepresented in the military (in the Army in particular), the military's success in recruiting qualified black women from a relatively small pool merits further inquiry. Young white women have the best results—about half meet the weight standards.

The results shown in Figure 4.1 are consistent with the results using the NLSY, shown in Figures 3.10 and 3.11. For example, Figure 4.1 shows that about 80 percent of white males would qualify for service (or about 20 percent would not qualify) and 70 to 75 percent of black

Figure 4.1
Percentage of Males Failing to Meet Weight Standards of at Least One Branch

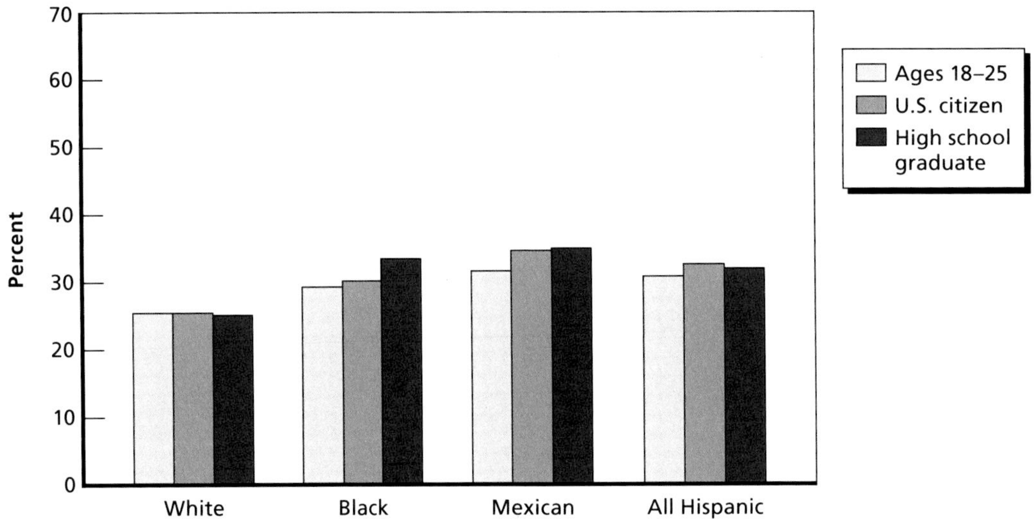

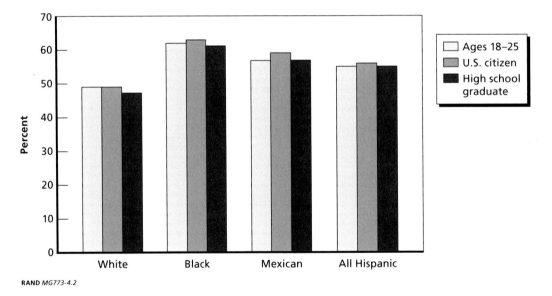

RAND *MG773-4.1*

Figure 4.2
Percentage of Females Failing to Meet Weight Standards of at Least One Branch

RAND *MG773-4.2*

males and Hispanic males would qualify (or about 25 to 35 percent would not qualify)—consistent with Figure 3.10. A similar consistency is found for females, especially relative to the results for the Army.

Figure 4.3 shows how the picture changes when we use the Air Force weight standards. The Air Force has the least restrictive weight standards, although it has more-difficult standards in other domains. Using these standards, Mexicans do not appear as disadvantaged,

Figure 4.3
Percentage of Males Failing to Meet Military Weight Standards of the Air Force
(the Least Strict)

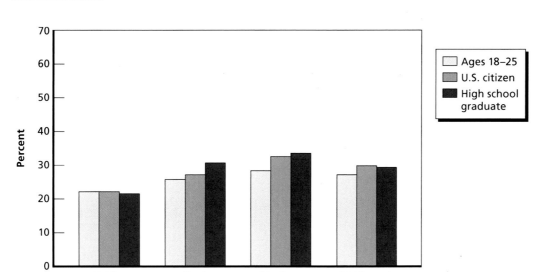

RAND MG773-4.3

showing a similar profile to blacks. However, it is noteworthy that even with relaxed weight criteria, approximately 30 percent of black and Hispanic (including Mexican) young men who are both citizens and high school graduates would not meet the weight requirements for military service.

Figure 4.4 shows that nearly 50 percent of all young black women and approximately 40 percent of all young Mexican women would fail the Air Force weight standards. White women fare better, with only about 30 percent failing the Air Force weight standards.

Figure 4.4
Percentage of Females Failing to Meet Military Weight Standards of the Air Force
(the Least Strict)

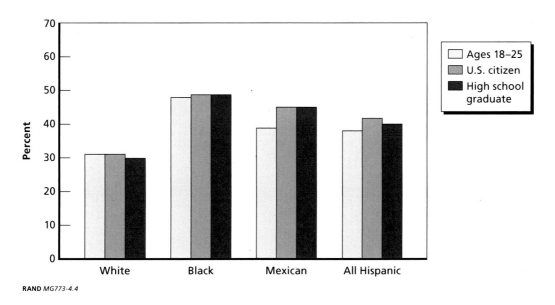

RAND MG773-4.4

Unlike the results for weight, Figures 4.5 and 4.6 show that Mexicans appear to have the lowest prevalence of major disqualifying health conditions. The effect of the selection criteria is different for whites, blacks and Mexicans. For Mexicans (even more so for all Hispanics), being a citizen has a large effect (especially among men), with citizens being *less* healthy than the general population of young Mexican men. For blacks, male high school graduates appear to be slightly less healthy than the average young black man. Young women of all ethnic groups appear to be less healthy than young men.

Figure 4.5
Percentage of Males with at Least One Major Disqualifying Health Condition

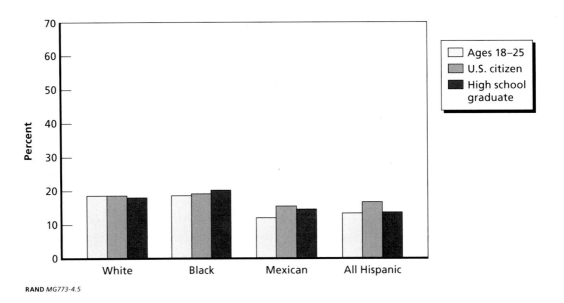

RAND *MG773-4.5*

Figure 4.6
Percentage of Females with at Least One Major Disqualifying Health Condition

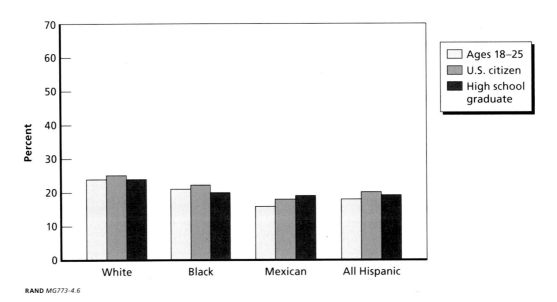

RAND *MG773-4.6*

Because of the uncertainty concerning how minor conditions may be treated at enlistment (e.g., they may be waiverable or may not be seen as a problem at all), Figures 4.7 and 4.8 aggregate two major disqualifiers: weight and the presence of a major condition. They thus capture the prevalence of nonwaiverable health conditions. We see that, although Mexicans have a lower prevalence of major health conditions, their higher prevalence of weight disqualification

Figure 4.7
Percentage of Males with at Least One Nonwaiverable Disqualifying Health Condition (Major Condition or Weight Disqualification)

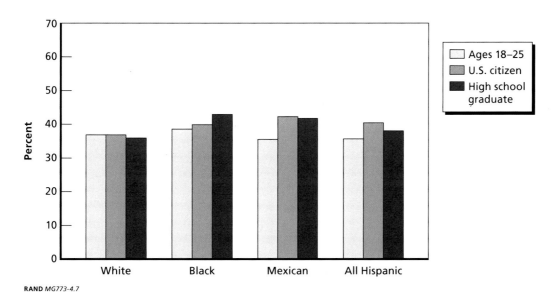

RAND *MG773-4.7*

Figure 4.8
Percentage of Females with at Least One Nonwaiverable Disqualifying Health Condition (Major Condition or Weight Disqualification)

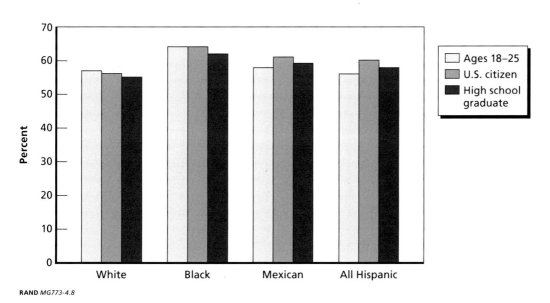

RAND *MG773-4.8*

appears to act as a counter, resulting in a prevalence of nonwaiverable conditions similar to that of blacks (if not slightly higher). As with previous patterns, we see that limiting the analysis to citizens significantly raises prevalence rates among Mexicans. The results for young women are dramatic, showing that over 60 percent of black women and over 55 percent of Hispanic women have a nonwaiverable health condition.

Finally, Figures 4.9 and 4.10 consider the net effect of health, identifying whether an individual has at least one of the three kinds of disqualifying health factors (weight, major

Figure 4.9
Percentage of Males with at Least One Disqualifying Health Condition (Major, Minor, or Weight Disqualification)

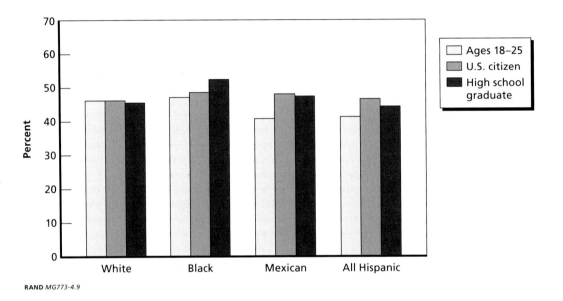

Figure 4.10
Percentage of Females with at Least One Disqualifying Health Condition (Major, Minor, or Weight Disqualification)

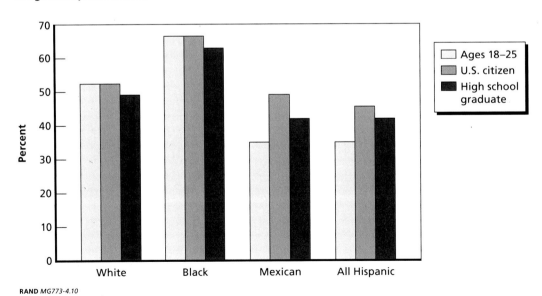

condition, or minor condition). Again, because of the higher prevalence of weight problems among Mexicans but the lower prevalence of other health conditions, there is little net difference across ethnic groups when aggregating these measures. But the effect for women is striking. Across all ethnic groups, women have poorer health profiles than men. Black women suffer from the poorest health: Nearly 70 percent have a health condition that may disqualify them from military service.

Discussion

The prevalence estimates using recent health data are revealing. It is clear that Hispanic men face obstacles in meeting the weight standards for each of the branches of the armed forces, with 32 percent of the probable military population of Mexican origin men and 31 percent of all Hispanic men not meeting the weight standards for new recruits of at least one service (compared with 25 percent of whites and 29 percent of blacks, as shown in Figure 4.1). On other health measures, all Hispanic men—especially men of Mexican origin—appear to have better health profiles than their white and black peers. Though restricting the population to the probable military population reduces the health advantage of Hispanic men, they still appear less likely to suffer from a major health condition, even after controlling for having seen a doctor in the past year. If major, minor, and weight conditions are considered together, about 45 percent of the probable military population would not qualify for military service.

A slightly different picture emerges for adult women. Like their male peers, Mexican and Hispanic women are less likely to have a major or minor disqualifying health condition. They are, however, more likely to fail the weight standards.

Although weight is a challenge for all groups, black women have the poorest weight profile. More than 60 percent of black women would fail the standards of at least one branch. Patterns of selection in general mirror those found in the male sample, with a few slight differences. The most striking difference concerns the prevalence of health problems in the female population. The presence of disqualifying conditions among females was greater than that among males by about 20 percentage points.

Given the striking disqualified weight profile for Hispanic men and black women, we conducted some additional analyses to evaluate the distribution of individuals around the weight cutoff points. That is, are there a large number of individuals within a few pounds of the weight standards, and does this distribution differ by race/ethnicity? If so, a slight relaxation of weight standards could result in more individuals being eligible for service. Appendix E presents kernel density plots of the difference between observed weight and the maximum allowable new accession weight, by race/ethnicity. These density plots can be usefully summarized by examining the percentage of individuals who would fail the military new accession weight standards and are over the maximum weight by five pounds or less and by ten pounds or less, by race/ethnicity (Tables 4.1 and 4.2).

Whites and all Hispanics are the most likely to be overweight by five pounds or less (conditional on being over the maximum standard weight), while all Hispanics and Mexicans are the most likely to be overweight by ten pounds or less (conditional on being over the maximum standard weight). Unconditional results show what percentage of the respective ethnic group would become eligible if weight standards were relaxed by five pounds or by ten pounds.

Table 4.1
Percentage of Males and Females Overweight Within Five Pounds and Ten Pounds of Maximum Allowable Weight, Conditional on Being Overweight

	White	Black	Mexican	Hispanic
Males				
Overweight 5 lbs or less	15.58	14.88	15.62	18.52
Overweight 10 lbs or less	29.68	25.65	35.52	36.78
Females				
Overweight 5 lbs or less	14.99	11.24	13.80	13.81
Overweight 10 lbs or less	26.72	22.65	27.23	26.92

Table 4.2
Percentage of Males and Females Overweight Within Five Pounds and Ten Pounds of Maximum Allowable Weight, Unconditional

	White	Black	Mexican	Hispanic
Males				
Overweight 5 lbs or less	4.23	5.17	5.42	6.21
Overweight 10 lbs or less	8.07	8.91	12.33	12.33
Females				
Overweight 5 lbs or less	5.73	6.52	7.30	6.78
Overweight 10 lbs or less	10.22	13.13	14.40	13.21

However, changing one standard in isolation does not necessarily result in a proportional change in the pool of eligible candidates. Standards tend to be correlated. For example, many candidates who are significantly overweight also lack a high school diploma and have lower AFQT scores. See Chapter Three for a discussion of how changing one or more standards is likely to affect eligibility.

Summary of Descriptive Results

The analyses show that noncitizen Mexican-origin males of all ages are healthier than their citizen peers, lending credence to the previously documented Hispanic paradox. (However, the same pattern of difference is not observed among females.) This pattern of healthier noncitizens is found for virtually every Hispanic ethnic group except Cubans and Puerto Ricans. It is unclear what circumstances would lead to Puerto Ricans being noncitizens, and these few cases may be erroneous or especially unusual cases. Cuban immigrants also do not show consistent evidence of being healthier noncitizens. This may be the result of the changing human and social capital of recent Cuban immigrants relative to the earlier Cuban refugees and their children born in the United States. That is, recent immigrant noncitizens have much lower human and social capital than the early waves of Cuban immigrants, whose children are likely to be in the citizen comparison group in this age range. That these results do not hold for females raises questions about differences in selection and possibly about behaviors between genders.

Summary of Findings and Conclusions

For convenience, Table 4.3 presents a summary of the findings from the descriptive analyses of the population samples. On balance, Hispanics appear to be equally healthy as whites, if not healthier in some areas. However, there is one major exception: Weight is an area in which Hispanics, especially Mexicans, face significant barriers to enlistment.

Table 4.3
Summary of Estimated Percentage of Young Adult Men and Women with a Disqualifying Health Condition, by Race/Ethnicity

	Weight	Air Force Weight	Major Condition	Nonwaiverable Condition[a]	Any Disqualifying Condition
Males					
White	25–28	21–25	18–21	36–41	46–57
Black	29–35	26–32	19–21	39–44	47–56
Mexican	32–37	27–32	12–16	36–42	41–51
Hispanic	31–36	27–30	13–17	35–41	41–51
Females					
White	47–49	30–33	23–25	55–57	64–69
Black	61–64	48–52	20–25	62–67	67–75
Mexican	57–59	39–47	16–19	58–67	60–66
Hispanic	53–56	38–42	18–20	56–60	60–63

[a] Nonwaiverable conditions are failing weight standards and major health conditions.

CHAPTER FIVE

Analysis of Hispanic Military Career Outcomes

The services screen applicants to ensure that recruits meet enlistment standards and that enlistees are productive members of the military. Yet these standards can be obstacles to entry for many youth, particularly minorities. As discussed in earlier chapters, high school graduation, AFQT, and weight are the key standards that disproportionately screen out Hispanic and black youth relative to white youth.

Several types of policy changes would have the effect of increasing minority recruitment and representation in the military. Chapter Six presents analysis relevant to the topic of recruiting more intensively among minorities who already meet the military's entry requirements. This chapter presents analysis relevant to the topic of relaxing standards—specifically those standards that tend to be the most likely to disqualify Hispanics, namely lack of high school diploma, lower AFQT score, and weight that exceeds the weight standard. Specifically, we examine the performance of minorities in the military, measured in terms of career outcomes (retention and promotion), and analyze how waivers and variations in standards affect these outcomes. The analysis addresses the following questions:

- How does varying quality (AFQT, high school graduation status) affect career outcomes?
- To what extent do career outcomes for minority and white recruits differ?
- To what extent do outcomes for lower quality minorities differ from those of similar white recruits?

Rather than explicitly examining the effects of waivers on career outcomes, the analysis examines how varying the standards affects career outcomes. This approach permits us to incorporate into the analysis the effects on outcomes for those who did not require a waiver. Thus, our data include individuals who met the standards, and those who required a waiver. By including both types of individuals, the analysis generalizes to all recruits, captures variation in career outcomes among all recruits, and assesses the effects of differences in entry characteristics on subsequent career outcomes for all recruits.

The chapter proceeds as follows. The first three sections present a conceptual framework, a review of past studies, and a description of our data. The following three sections present simple comparisons across race/ethnic groups, our regression methodology, and our regression-corrected estimates. The body of the chapter presents our empirical results. The chapter concludes with a brief discussion of the implications of the findings.

Conceptual Framework

What factors would we expect to affect promotion and retention? We consider first the simpler case of promotion and then the more complicated case of retention. In each case, we consider first the effect of measures of service member quality and then the pure effect of race/ethnicity.

Early Attrition and Promotion

In general, we expect that attrition and promotion are functions solely of individuals' performance in the military, rather than of race or ethnicity. We would expect people who have better AFQT scores and more civilian education and who are lighter (relative to their height) and older (presumably more "mature") to perform better in the military. Those who perform better are less likely to wash out of basic training (which we measure using three-month retention) and more likely to be promoted early (which we measure using promotion to E-5 by four years) or on time (which we measure using promotion to E-5 by six years).

There are two possible explanations for any remaining effect of race/ethnicity. First, one could argue that based on their racial characteristics, whites (blacks and Hispanics) are more (less) likely to be promoted. We will see that the evidence is exactly the opposite.

Second, such an effect might be because race/ethnicity is a proxy for other (unmeasured to the analyst) characteristics. Such characteristics might include physical condition, "drive," and the ability to follow instructions and deal with adversity. The observed characteristics—AFQT, education, weight, age—are likely to be correlated with these unmeasured characteristics, but they are unlikely to perfectly explain these characteristics. Thus, for example, if—given their observed characteristics—blacks or Hispanics had better physical condition (relative to whites) on average, then we might estimate that being black or Hispanic would have a positive effect on promotion.

As we discuss in the conclusion to this chapter, inasmuch as this unmeasured-characteristic explanation is correct, DoD might want to consider it in making recruiting decisions. Thus, knowing that—for given observable characteristics—black or Hispanic recruits are more likely to be promoted, we might reasonably infer that they were therefore (on average) better service members. In that case, it might be cost-effective to devote relatively more recruiting resources to those groups. Such recruiting resources might include advertising dollars, assignment of recruiters, and guidance to recruiters about which strategies to use and which prospective recruits to work on more intensively.

Retention

Analysis of the expected effect of standards on retention is more subtle. The standard economic model of retention (e.g., Gotz and McCall, 1984; Asch and Warner, 1994) posits that in making reenlistment decisions, service members compare their projected future military earnings with projected future civilian earnings if they leave the military. (The term *earnings* is broadly defined and includes both cash and nonpecuniary factors, such as enjoyment of the military lifestyle.) In the military, earnings rise with promotion speed. Any factor that leads to faster promotion (see the discussion in the previous subsection) will lead to higher expected military earnings. In addition, it seems plausible that those who are promoted faster will derive more nonpecuniary satisfaction from military service. This might be due to some combination of the facts that the military is more psychically rewarding at higher ranks (e.g., more responsibility, more discretion, more leadership opportunities) and that people who derive more satisfaction from the military may be more likely to work harder at it and thereby be promoted.

However, promotion and satisfaction with the military are not the only factors in the reenlistment decision. The standard economic model posits that service members compare their military earnings to what they project they would earn as civilians. Many of the factors that increase the probability of promotion and therefore military earnings also increase likely civilian earnings. The civilian labor market pays better-educated people more, and civilian earnings rise with AFQT. Though the military pay table does not depend explicitly on either AFQT or education, promotion speed depends on both, as do qualification for enlistment bonuses, better education benefits such as the Army College Fund, and some special pay rates targeted to occupations requiring higher aptitude scores. Thus, we cannot project the effect of AFQT score, education, and other factors (such as weight) on retention.

The analysis for race/ethnicity itself is less clear. The conventional analysis has assumed that the military is closer to being a pure meritocracy than the civilian labor market is. The military is also probably closer to being "color blind" and "ethnicity blind." Family connections probably matter less. Inasmuch as the civilian labor market is less of a meritocracy—e.g., prejudice against blacks and Hispanics remains; family connections matter more in the labor market and minorities have weaker connections—blacks and Hispanics will have lower civilian earnings than otherwise identical whites. This will cause the military to look relatively more attractive to minorities. That is the conventional explanation of the higher-than-expected enlistment rates of blacks.

It should be noted, however, that the existence and magnitude of any such pure race/ethnicity effect is the subject of some controversy in the academic literature. It is indisputable that, in the civilian labor market, blacks and Hispanics have lower earnings than whites. However, much of that difference is explained by lower education, lower AFQT scores, and (for Hispanics) poorer English language skills (often because of immigrant status). Controlling for such variables eliminates some (and occasionally all) of the pure race/ethnicity effect (e.g., Hotz et al., 2002; see the review in Duncan, Hotz, and Trejo, 2006; for blacks, see Neal and Johnson, 1996).

Further study on the pure effect of race and ethnicity on civilian earnings is needed. However, it seems reasonable to conclude that some earnings penalty exists. This conclusion is reinforced by the fact that wage studies have not incorporated wage observations for nonworkers (see the discussions in Neal and Johnson, 1996, and Hotz et al., 2002). Such nonworkers are more common among blacks and Hispanics. Controlling for this varying percentage of nonworkers is likely to increase the magnitude of the effect of race and ethnicity.

Past Studies

Not surprisingly, given their importance to the military, retention and promotion have been standard topics in the military manpower literature. In that literature, race/ethnicity has been included, but only as a control variable. Here, we survey the existing literature, extracting the estimates of race/ethnicity effects.

For the Army, Buddin (2005) explored the effects of recruit characteristics and other factors on the success of first-term soldiers. His data covers the period 1995 to 2001. He found that, other factors constant, Hispanic and black Army recruits had substantially lower basic combat training (BCT) attrition rates (i.e., higher retention rates). For other covariates, he found that AFQT had a small but negative effect on BCT attrition, while those with a GED had substantially higher attrition rates. He also estimated a joint model of reenlistment and promotion and found qualitatively similar effects of these variables on the reenlistment decision. Hispanic and black soldiers were promoted faster to E-5, all else being equal. Faster promotion was also seen for those with higher AFQT scores and with some college.

For the Marine Corps, Hattiangadi, Lee, and Quester (2004) examined boot camp and first-term attrition rates. Their data cover the period 1979 to 2001. They found that attrition rates were lower (i.e., retention rates were higher) among Hispanic recruits relative to other race/ethnic groups. This pattern held for both males and females. On the other hand, they found relatively small differences in the marginal effects of other characteristics (e.g., AFQT, citizenship status) on attrition for Hispanics and non-Hispanics. Also, except for citizenship and boot camp location (Hispanic men are less likely to train at Parris Island), there are relatively few differences in the characteristics of Hispanic and non-Hispanic Marine recruits. Thus, the relatively lower attrition rates of Hispanic Marines appear to be due to unobserved characteristics. Fieldwork by the authors suggested that discipline and other family-related variables may explain the difference. In a follow-up study, Hattiangadi et al. (2005) found that noncitizens had lower three-month attrition and 36-month attrition than did citizens with similar characteristics.

A number of past studies provide evidence of the effects of AFQT, education and metrics of personnel quality on retention and promotion. Recent studies include Asch, Romley and Totten (2005) and Hosek and Mattock (2003). Consistent with the conceptual framework outlined earlier (i.e., higher-quality personnel have good opportunities both within the military and in the civilian sector), these studies found that the average AFQT of those who attrite is not much different from the average AFQT of those who complete their first term, that those who reenlist are of slightly lower quality (in terms of AFQT) than those who separate at the end of the first term, and that those who stay until their early and mid-careers are not much different in terms of average AFQT than those who leave. However, through their mid-career, those who are promoted have significantly higher AFQT scores than those who are not. The relatively small effect of AFQT on retention is similar to that found in Buddin (2005) and in earlier studies (Warner and Solon, 1991, Smith, Sylwester and Villa, 1991). Consistent with Buddin (2005), earlier studies found that lack of a high school diploma increased attrition (Warner and Solon, 1991).

Data

In the balance of this chapter, we estimate regression models of the effect of race/ethnicity on career outcomes and how those effects vary with observable characteristics. Specifically, we consider the determinants of five outcomes:

1. **Retention at Three Months.** This measure captures whether recruits survived boot camp and initial occupational training.
2. **Retention at Four Years of Service.** This measure approximates completion of the first enlistment term for most service members.

3. **Retention at Six Years of Service.** This measures approximate completion of the first enlistment term for technical skills in the Air Force and Navy, both of which have enlistment terms as long as six years.

4. **Achieving Pay Grade E-5 by Four Years of Service.** This promotion speed outcome approximates early promotion.

5. **Achieving Pay Grade E-5 by Six Years of Service.** This promotion speed outcome approximates on-time promotion.

We explore these issues using data generously provided by John Warner of Clemson University. That file contains a longitudinal record of military careers created by merging a subset of variables from the annual enlisted active duty contract transaction records and master file records for all enlisted personnel entering military service between September 30, 1988, and September 30, 2002. The file was constructed as part of an ongoing Clemson project on military recruiting and subsequent career outcomes. These records were provided by the Defense Manpower Data Center (DMDC), and permission to use the data for this project was given to RAND. To these longitudinal files, Warner appended variables describing the area (for most variables, this was the county) from which the recruit accessed (e.g., unemployment rate, percentage of veterans) and characteristics of the enlistment contract (e.g., length of contract, receipt of education benefits, receipt of cash bonuses).

To these data, we also appended information from the USMEPCOM examination accession files. These data provided detailed information on military qualification characteristics (e.g., AFQT score, education, weight). The MEPCOM data were provided to RAND by DMDC.

The final analysis file had 569,954 observations. The original file had an additional 12,769 observations that were dropped because of missing values for the covariates (usually education).

Career Outcomes for Hispanics Versus Other Races and Ethnicities

Table 5.1 reports the mean values across the accession cohorts of the five outcomes for each service, by race/ethnicity, as well as the differences for Hispanics and blacks in outcomes relative to white recruits. Two key results are apparent. First, relative to white recruits, both black and Hispanic recruits are more likely to stay in service and also more likely to be promoted. For example, on average 88 percent of white Army recruits are retained in the first three months while 92 percent of Hispanic and black recruits are retained. In the case of promotion, 17 percent of white E-5s are promoted in the Army by the sixth year of service while 22 percent of Hispanic recruits and 23 percent of black recruits are promoted by then.

The second result is that the gaps between minority and white outcomes in both retention and promotion grow with years of service. The gap between white and Hispanic recruits in the three-month Army retention rate is 4 percent (92 percent – 88 percent). This gap grows to 7 percent (46 percent minus 39 percent) in the four-year and six-year (33 percent – 26 percent) retention rate. Similarly, the gap between black and white Army recruits grows from 4 percent at 3 months to 11 percent at six years. Similar results are found for the other services, although they are less dramatic. For example, in the Air Force, the gap between Hispanic and white recruits in three-month retention rates is 4 percent; at six years it is 5 percent. The findings of

Table 5.1
Raw Outcomes, by Service and Race/Ethnicity

Outcome	Rates (%)			Difference Relative to Whites (%)	
	White	Black	Hispanic	Black	Hispanic
Army					
Retained at 3 months	88	92	92	4	4
Retained at 4 years	39	49	46	9	7
Retained at 6 years	26	37	33	11	7
Promoted to E-5 at YOS6	17	23	22	6	5
Promoted to E-5 at YOS4	11	14	14	3	3
Navy					
Retained at 3 months	85	87	87	3	2
Retained at 4 years	46	50	50	4	4
Retained at 6 years	27	35	29	8	2
Promoted to E-5 at YOS 6	22	22	21	0	0
Promoted to E-5 at YOS 4	15	13	14	−2	−1
Marine Corps					
Retained at 3 months	86	87	89	1	3
Retained at 4 years	59	61	67	2	8
Retained at 6 years	11	18	17	7	6
Promoted to E-5 at YOS 6	13	15	18	2	5
Promoted to E-5 at YOS 4	8	7	11	−1	3
Air Force					
Retained at 3 months	93	97	97	4	4
Retained at 4 years	78	80	83	2	5
Retained at 6 years	57	66	61	10	4
Promoted to E-5 at YOS 6	32	35	34	2	2
Promoted to E-5 at YOS 4	3	3	3	0	0

NOTE: YOS = year of service.

higher retention rates and lower attrition rates for minorities are similar to those discussed in our review of previous studies.

A key question motivating our analysis is whether minorities who were heavier, had lower AFQT, or had less education continued to have outcomes that exceed those of white recruits. The retention and promotion rates presented in Table 5.1 control only for contract year. They do not control for differences in characteristics that might affect outcomes, such as enlistment contract terms, characteristics of the county where the recruit enlisted, and personal characteristics of the recruit. To control for such observable characteristics, in the next subsection we employ a regression methodology to estimate the effect of race/ethnicity for otherwise observationally similar individuals.

Linear Regression Methodology

Our regression specification is flexible. We estimate separate models for each outcome, for each service. Furthermore, because our primary interest is in race/ethnic group effects, our regression models interact each of the covariates with race/ethnicity.

Specifically, for each outcome and each service, we estimate a linear regression model (i.e., the linear probability model). There are thus 20 regressions in total (i.e., five outcomes for each of four services). Each of the models has the form

$$y_i = \alpha + r_i{}'\rho + X_i\beta + r_i X_i \gamma + \varepsilon_i \qquad (5.1)$$

where y is one of the five outcomes of interest, r is a vector of dummy variables identifying blacks and Hispanics and X is a vector of other covariates (individual characteristics, local characteristics, and contract terms). The first term, α, is a constant. The second term is the main effect for race ethnicity, relative to whites, with coefficients ρ. The third term is the simple controls for covariates, with coefficient β. The fourth term is a complete set of interactions between race/ethnicity and the other covariates, with coefficients γ. The final term, ε, is the unexplained residual.

The variables included in X, the characteristics related to entry standards, include AFQT, education, weight (relative to service-specific standards), number of dependents, marital status, being a noncitizen, and age. Other control variables in X include gender (a personal characteristic not related to enlistment standards) and enlistment contract terms (years of service obligation, an indicator of the college fund, the nominal dollar amount of any enlistment bonus), calendar year, and county characteristics (fraction of the county's male population ages 35 and older who are veterans, the county's per capita income, and populate density per square mile). The rationale for including county variables is that recruits from locations with larger populations of veterans have a stronger military influence and a stronger propensity to serve in the military. Those from wealthier locations have better civilian opportunities and are less likely to stay in service; those from locations with denser populations are less likely to enlist. All variables are measured as of the contract date.

Table 5.2 shows the mean value of the variables as well as the mean values by race and ethnicity. Relative to their distribution in the services overall, blacks are overrepresented in the Army and slightly underrepresented in the Marines and Air Force. In contrast, Hispanics are underrepresented in the Army and Air Force, and overrepresented in the Navy and the Marines.

The AFQT distribution is radically different across the races. Whites are overrepresented in Categories I and II, while blacks and Hispanics are underrepresented. Representation in Category IIIA is approximately constant, and representation in Category IIIB shows the mirror image of Categories I and II. There are very few enlistees at Category IV or below. There is little pattern in the education distribution.

Consistent with the evidence on the population as a whole, whites are more likely to be in the highest group. Hispanics are more likely to be above the standard.

Table 5.2
Sample Means for Covariates, Overall and by Race/Ethnicity (%)

	Overall	White	Black	Hispanic
Race/ethnicity	100	71	19	10
Army	36	34	43	32
Navy	27	26	27	30
Marines	19	20	13	26
Air Force	18	20	16	12
Cat I or II	41	48	21	28
Cat IIIA	28	28	29	30
Cat IIIB	30	24	48	41
Cat IV or V	1	1	1	1
High school dropout	0	0	0	0
In high school	36	36	33	36
High school graduate	53	52	58	53
Some college	5	5	5	5
Alternative high school	5	6	3	5
Weight 31+ below	42	44	38	36
Weight 21–30 below	15	15	16	16
Weight 11–20 below	14	13	14	15
Weight 1–10 below	12	12	12	14
Weight 0–4 above	5	5	5	6
Weight 5–9 above	4	4	4	5
Weight 10–19 above	5	4	5	5
Weight 20+ above	3	3	5	3
Term 2 years	2	2	1	2
Term 3 years	16	15	22	18
Term 6 years	9	10	5	6
Term other (4 years)	73	74	71	74
Female	18	16	26	16
Married	6	6	5	7
Any dependents	5	4	5	5
Noncitizen	4	2	5	16
Age at contract	19.7	19.7	19.8	19.7
College Fund	13	15	9	11
Any bonus	4	4	2	3
Bonus amount	$169	$202	$71	$118
County characteristics				
Unemployment rate	5	5	5	6
Per-capita income ($ thousands)	20	21	20	20
Number of veterans	64	63	63	67
Population density (per square mile)	10	9	14	9
Percent college	37	37	36	34

Whites are more likely to enlist for six years. Blacks (and to a lesser extent Hispanics) are more likely to enlist for only three years.

Blacks are more likely to be female. Not surprisingly, Hispanics are more likely to be non-citizens. There is no pattern in age at contract.

Whites are more likely to sign up for the College Fund, more likely to get a bonus, and likely to get a larger bonus.

Linear Regression Results

We estimated Equation (5.1) using the accession cohort data. We then used the regression model to predict outcome measures by race/ethnicity and to predict outcomes by different values of the entry standard characteristics, given race/ethnicity. The analysis focuses on the three entry standard characteristics, namely, AFQT, education, and weight. As we saw earlier, these are the key disqualifying characteristics of Hispanic youth.

Individual regression coefficients are difficult to interpret. Here, we report predictions from these regression models, holding the other observable characteristics at their mean values (over all race/ethnicities; not within a race/ethnicity).

Adjusted Versus Raw Outcomes, by Race and Ethnicity

Table 5.3 compares predicted outcome measures by race/ethnicity, adjusting for the personal characteristics, contract terms, and county characteristics included in the regression model. Specifically, for both the raw and regression adjusted estimates, it reports the outcome for blacks and Hispanics relative to whites (i.e., black/Hispanic minus white). The table suggests that regression adjustment makes little difference for Hispanics. Comparing column (8), the raw differences between white and Hispanic outcomes to column (10), the adjusted differences, differences tend to be small. Put differently, regression adjustment leads to inferences similar to the raw differences presented in Table 5.1.

For example, for the Army, the raw four-year retention rate is 7 percent, while the adjusted rate is 6 percent. Similarly, the adjusted and raw three-month retention rates are the same, namely 4 percent. Again in the case of the Air Force, the differences relative to white recruits are virtually the same. The raw and adjusted difference between Hispanic and white recruits in four-year retention rate is 5 percentage points. Thus, even though some observable characteristics differ between Hispanic and white recruits, the net effect of controlling for observable characteristics is relatively small.

The substantive implication is that, as in the case of the raw outcomes, the differences in the adjusted outcomes are generally positive. That is, Hispanics generally have higher retention rates and faster promotion, both before and after we control for observable characteristics. The exception is promotion in the Navy, where we find negative differences in the adjusted rates between Hispanic and white recruits.

The similarity is not as great between the adjusted and the raw differences for black recruits. For them, in most cases, controlling for observable characteristics shrinks the effect of race. For example, in the case of the Army, the raw difference in four-year retention rates between white and black recruits is 9 percentage points, but the adjusted difference is 6 percentage points. In other cases, the differences are more similar, as in the case of the differences in six-year Army retention rates.

Table 5.3
Raw and Adjusted Outcomes, by Race/Ethnicity (%)

| Outcome | Raw Rates | | | Adjusted Rates | | | Raw Differences Relative to Whites | | Adjusted Differences Relative to Whites | |
	White (1)	Black (2)	Hispanic (3)	White (4)	Black (5)	Hispanic (6)	Black (7)	Hispanic (8)	Black (9)	Hispanic (10)
Army										
Retained at 3 months	88	92	92	91	93	95	4	4	2	4
Retained at 4 years	39	49	46	48	54	54	9	7	6	6
Retained at 6 years	26	37	33	26	36	34	11	7	10	8
Promoted to E-5 at YOS6	17	23	22	20	24	25	6	5	4	5
Promoted to E-5 at YOS4	11	14	14	15	15	17	3	3	0	3
Navy										
Retained at 3 months	85	87	87	88	89	90	3	2	1	3
Retained at 4 years	46	50	50	60	59	64	4	4	−1	4
Retained at 6 years	27	35	29	31	33	32	8	2	2	1
Promoted to E-5 at YOS6	22	22	21	28	20	25	0	0	−8	−3
Promoted to E-5 at YOS4	15	13	14	20	11	16	−2	−1	−9	−5
Marine Corps										
Retained at 3 months	86	87	89	89	90	93	1	3	0	3
Retained at 4 years	59	61	67	66	66	76	2	8	0	10
Retained at 6 years	11	18	17	16	22	21	7	6	6	5
Promoted to E-5 at YOS6	13	15	18	21	21	24	2	5	0	4
Promoted to E-5 at YOS4	8	7	11	13	10	14	−1	3	−3	1
Air Force										
Retained at 3 months	93	97	97	91	93	95	4	4	2	4
Retained at 4 years	78	80	83	76	76	81	2	5	0	5
Retained at 6 years	57	66	61	41	48	45	10	4	8	5
Promoted to E-5 at YOS6	32	35	34	25	25	27	2	2	0	2
Promoted to E-5 at YOS4	3	3	3	2	1	2	0	0	−1	0

Effects of Entry Standard Characteristics on Outcomes, by Race and Ethnicity

The previous subsection considered the total effect of race/ethnicity. In this subsection, we consider the effect of the three key entry standards. Specifically, Tables 5.4 through 5.9 present the model's predictions of how outcomes change as AFQT, education and weight, respectively. In each table, the predictions are computed within race/ethnicity group, holding constant all other observable characteristics at their grand mean values across all race/ethnic groups.

Table 5.4
Predicted Outcomes, by AFQT Category and Race/Ethnicity

Service	Outcome	AFQT Category	Percent		
			White (1)	Black (2)	Hispanic (3)
Army	Retained at 3 months	I–II	90	93	92
		IIIA	88	92	92
		IIIB–IV	86	92	92
	Retained at 4 years	I–II	42	47	49
		IIIA	38	45	46
		IIIB–IV	38	45	50
	Retained at 6 years	I–II	27	34	33
		IIIA	26	34	32
		IIIB–IV	26	34	35
	Promoted to E-5 at YOS6	I–II	20	24	24
		IIIA	17	21	22
		IIIB–IV	14	18	21
	Promoted to E-5 at YOS4	I–II	15	16	17
		IIIA	10	13	15
		IIIB–IV	7	11	14
Navy	Retained at 3 months	I–II	87	88	89
		IIIA	84	86	87
		IIIB–IV	82	85	86
	Retained at 4 years	I–II	49	49	50
		IIIA	45	48	49
		IIIB–IV	43	48	49
	Retained at 6 years	I–II	30	31	31
		IIIA	26	31	29
		IIIB–IV	26	31	27

Table 5.4—Continued

Service	Outcome	AFQT Category	Percent White (1)	Black (2)	Hispanic (3)
	Promoted to E-5 at YOS6	I–II	29	24	29
		IIIA	20	18	22
		IIIB–IV			
	Promoted to E-5 at YOS4	I–II	23	16	20
		IIIA	13	10	12
		IIIB–IV	8	6	8
Marine Corps	Retained at 3 months	I–II	88	89	90
		IIIA	86	87	90
		IIIB–IV	84	86	88
	Retained at 4 years	I–II	63	65	70
		IIIA	58	61	70
		IIIB–IV	55	59	68
	Retained at 6 years	I–II	14	21	22
		IIIA	11	16	17
		IIIB–IV	8	15	15
	Promoted to E-5 at YOS6	I–II	18	22	23
		IIIA	11	16	18
		IIIB–IV	8	14	16
	Promoted to E-5 at YOS4	I–II	12	11	13
		IIIA	7	8	10
		IIIB–IV	4	6	8
Air Force	Retained at 3 months	I–II	95	95	96
		IIIA	94	95	96
		IIIB–IV	93	94	95
	Retained at 4 years	I–II	80	82	83
		IIIA	77	81	81
		IIIB–IV	75	80	79
	Retained at 6 years	I–II	57	61	61
		IIIA	57	62	63
		IIIB–IV	57	62	65
	Promoted to E-5 at YOS6	I–II	36	34	34
		IIIA	32	31	30
		IIIB–IV	29	29	29
	Promoted to E-5 at YOS4	I–II	4	3	3
		IIIA	3	2	3
		IIIB–IV	3	2	2

We first consider column (1), the predicted retention outcomes for white recruits. We find a generally strong effect of AFQT on retention for white recruits. That is, as AFQT category rises, retention also rises. For example, the predicted four-year retention for white Marine Corps recruits who scored in AFQT categories I and II is 63 percent, but is eight percentage points lower, or 55 percent, for similar white Marine Corps recruits who scored in AFQT categories IIIB and IV. We see evidence of similar patterns in the other services, suggesting that higher-aptitude white recruits are more successful in the military, and that opportunities in the military are sufficiently attractive to induce them to stay rather than leave.

For Hispanic recruits, the effect of AFQT on retention is less clear-cut, though generally, the effect is weakly positive. For example, in the case of the Marine Corps, the four-year retention rate of category I and II Hispanic recruits is 70 percent, two percentage points higher than the 68 percent retention rate for Hispanic recruits in AFQT categories IIIB and IV. Similarly, in the Navy, the adjusted four-year retention rate for Hispanic AFQT category I and II recruits is 50 percent, one percentage point higher than the adjusted rate for similar Hispanic recruits in AFQT categories IIIB and IV. Similar results are generally found for other retention outcomes and other services. The exception is the Army, where we find little discernable pattern in retention rates by AFQT for Hispanic recruits. In the case of three-month retention, adjusted rates for Hispanic recruits do not vary by AFQT category. Four-year retention rates decrease, then increase as AFQT increases, and in the case of six-year rates, retention improves as AFQT falls. These results are consistent with the conjecture that—because of language issues—the AFQT is not as good a predictor of military careers for Hispanics.

The results for black recruits are similar to those for Hispanic recruits. Retention rates vary positively but weakly with AFQT category. In the Marine Corps, black recruits in AFQT categories I and II have four-year retention rates that are six percentage points higher than similar recruits in AFQT categories IIIB and IV (65 percent versus 59 percent). They have higher three-month and six-month retention rates in the Marine Corps as well. Similar results are found in the Navy, Air Force, and Army. However, these other six-year retention rates for black recruits tend not to vary with AFQT. The general conclusion for both black and Hispanic recruits is that lower-quality personnel in terms of AFQT have slightly lower retention rates than higher-quality personnel. However, the effects of AFQT are smaller than are the effects for white recruits. These results are consistent with more-able recruits being more successful in the military and having better opportunities in the military than in the civilian sector.

These results of a positive effect of AFQT on promotion and retention—within race/ethnicity groups—are not surprising given the earlier literature. The insight from Table 5.4 comes from comparing across race/ethnic groups. The effects of race/ethnicity are generally larger than the effects of AFQT, and the magnitudes are large. Thus, the adjusted retention rate of white recruits in the group with the highest AFQT—categories I and II—is often lower than the retention rates of blacks or Hispanics in any AFQT group. For example, the adjusted four-year retention rate of white Army recruits in categories I and II is 42 percent, while the adjusted four-year retention rates of similar black and Hispanic recruits in AFQT categories IIIB and IV are 45 percent and 50 percent, respectively. In the case of the Air Force, the six-year retention rate of white recruits in category I and II is lower than the six-year retention rate of black or Hispanic recruits in any AFQT category. In summary, lower-quality minority recruits are more likely to remain in service than higher-quality white recruits with similar characteristics. These results suggest that a reduction in AFQT standards accompanied by an increased share of minority recruits would not result in a significant drop in retention—and might even result in an increase in retention.

In addition to predictions for retention rates, Table 5.4 also tabulates predicted promotion probabilities by AFQT category. As expected given our theory and the previous literature, the likelihood of promotion increases with AFQT. This result is consistent across all race/ethnic groups, though the results are often strongest for white recruits. For example, the adjusted probability that white Air Force recruits in AFQT categories I and II will be promoted to E-5 by YOS 6 is 36 percent and 29 percent for those in categories IIIB and IV. For the same categories, the promotion rates to E-5 at YOS 6 are 34 percent and 29 percent, respectively, for both black and Hispanic recruits. Thus, the promotion system is pro-selective in terms of AFQT. Since AFQT is correlated with performance on hands-on military-related tasks, the promotion system selects those who perform well and are well-suited for the military, regardless of race/ethnicity.

We saw that the effects of race/ethnicity on retention were larger than the effects of AFQT. That pattern carries over to promotion in some services, but not in others. For the Army, the race/ethnicity effects are nearly as large as the AFQT effects. For example, the adjusted six-year promotion rate for white recruits in AFQT categories I and II is 20 percent, one percentage point lower than the 21 percent adjusted promotion rate of Hispanic recruits in categories IIIB and IV. Thus, lower-quality Hispanic recruits are promoted faster than similar white recruits in higher AFQT categories. How is this possible? The Hispanic recruits have unobserved characteristics that result in faster promotion in the Army, despite their lower AFQT scores. However, in the Air Force and the Navy, the race/ethnicity effects are much smaller than the AFQT effects. For example, in the case of the Air Force, white recruits in categories I and II have a 36 percent six-year promotion rate to E-5, a rate that is higher than the six-year promotion rate to E-5 for either black or Hispanic recruits in any AFQT category.

Table 5.5 presents the results when the entry standard is education. Here, we consider how outcomes vary when education changes, by race/ethnicity, holding other observable characteristics constant. The results are qualitatively similar to those found in Table 5.4. First, we find, as did previous studies, that high school dropouts have lower retention rates, while those who are recruited while still in high school have higher retention rates, regardless of race or ethnicity. For example, the adjusted four-year retention rate for white Navy high school dropouts is 38 percent, but is 50 percent for high school graduates and 59 percent for those recruited while still in high school. Those in high school also have higher promotion rates to E-5 than high school dropouts or high school graduates. Continuing with the Navy example, 28 percent of Hispanics who are recruited into the Navy while still in high school are promoted to E-5 within six years, compared with 16 percent of high school dropouts with similar observable characteristics. Second, as with the results for AFQT, we generally find that the race/ethnicity effects nearly swamp the effects of education on adjusted outcomes. That is, white recruits who are still in high school have adjusted retention rates that are about the same or below the adjusted rates of minority dropouts, and white graduates have adjusted rates that are below the rates of minority dropouts. For example, the adjusted three-month Army retention rate of Hispanic dropouts is 90 percent, equal to the adjusted rate of whites recruited by the Army while still in high school. Again, the race/ethnicity effect generally swamps the education effect, except in the Navy. In the Navy, better-educated white recruits stay longer and are promoted faster than minority dropouts.

Table 5.5
Predicted Outcomes, by Education and Race/Ethnicity

Service	Outcome	Education When Recruited	Percent White (1)	Black (2)	Hispanic (3)
Army	Retained at 3 months	High school dropout	88	91	90
		High school graduate	88	93	93
		Still in high school	90	94	94
	Retained at 4 years	High school dropout	40	46	46
		High school graduate	39	49	48
		Still in high school	49	57	56
	Retained at 6 years	High school dropout	28	38	32
		High school graduate	27	37	35
		Still in high school	32	42	40
	Promoted to E-5 at YOS6	High school dropout	17	23	22
		High school graduate	17	23	23
		Still in high school	18	25	24
	Promoted to E-5 at YOS4	High school dropout	10	15	14
		High school graduate	11	14	15
		Still in high school	11	15	16
Navy	Retained at 3 months	High school dropout	81	84	82
		High school graduate	86	88	89
		Still in high school	90	90	93
	Retained at 4 years	High school dropout	38	43	40
		High school graduate	50	54	54
		Still in high school	59	60	61
	Retained at 6 years	High school dropout	26	29	21
		High school graduate	28	34	32
		Still in high school	36	39	37
	Promoted to E-5 at YOS6	High school dropout	20	17	16
		High school graduate	23	22	26
		Still in high school	28	24	28
	Promoted to E-5 at YOS4	High school dropout	11	9	8
		High school graduate	17	13	16
		Still in high school	20	14	18
Marine Corps	Retained at 3 months	High school dropout	87	89	91
		High school graduate	87	87	91
		Still in high school	89	90	91
	Retained at 4 years	High school dropout	62	68	72
		High school graduate	60	61	70
		Still in high school	67	70	74
	Retained at 6 years	High school dropout	11	21	18
		High school graduate	10	17	18
		Still in high school	16	22	21

Table 5.5—Continued

			Percent		
Service	Outcome	Education When Recruited	White (1)	Black (2)	Hispanic (3)
	Promoted to E-5 at YOS6	High school dropout	11	22	21
		High school graduate	16	18	22
		Still in high school	15	18	20
	Promoted to E-5 at YOS4	High school dropout	31	28	34
		High school graduate	32	31	33
		Still in high school	33	34	35
Air Force	Retained at 3 months	High school dropout	94	93	94
		High school graduate	93	96	96
		Still in high school	94	96	97
	Retained at 4 years	High school dropout	79	84	83
		High school graduate	78	80	82
		Still in high school	81	83	85
	Retained at 6 years	High school dropout	57	63	68
		High school graduate	56	62	63
		Still in high school	60	65	67
	Promoted to E-5 at YOS6	High school dropout	31	28	34
		High school graduate	32	31	33
		Still in high school	33	34	35
	Promoted to E-5 at YOS4	High school dropout	3	1	2
		High school graduate	2	2	2
		Still in high school	3	2	3

The third entry standard characteristic we consider is weight. As noted in Chapter One, weight is an important disqualifying characteristic for Hispanic and black youth. Tables 5.6 through 5.9 present the predicted outcomes for each service for weight, where weight is measured relative to the service-specific standard. As in Tables 5.4 and Table 5.5, these tables present predictions for the effect of race/ethnicity and weight, holding all other characteristics at their global mean values.

In general, those who are overweight, especially over 20 pounds overweight, have lower retention rates than those who are within five pounds (0 to 5) of the standard. This is especially true in the Army, the service with the most stringent weight standards. For example, Hispanic Army recruits who are over 20 pounds above the weight standard have an adjusted retention rate of 42 percent, compared with 49 percent for Hispanic recruits who are within five pounds of the standard. Those who are overweight have slower promotion rates as well. Thus, in the Navy, Hispanics who are over 20 pounds overweight have a promotion rate of 19 percent to E-5 within six years compared with 23 percent for those who are within five pounds of the standard.

Again, race/ethnicity effects dominate the weight effects; i.e., overweight minorities often have higher retention rates than white recruits who are near the weight standard. In the case of the Marine Corps, the differences are quite large. For example the four-year retention rate of a Hispanic Marine Corps recruit who is 20 pounds overweight is 69 percent, a rate that is 12 percentage points above the 57 percent four-year retention rate for white recruits who are

Table 5.6
Predicted Outcomes, by Weight Category and Race/Ethnicity, Army

Outcome	Weight Relative to Service Standard (lbs)	Percent		
		White (1)	Black (2)	Hispanic (3)
Retained at 3 months	Under 30+	89	93	93
	Under 20–30	89	94	93
	Under 10–20	89	93	94
	Under 0–10	88	92	91
	Over 0–5	88	92	92
	Over 5–10	88	92	91
	Over 10–20	88	92	91
	Over 20+	86	91	92
Retained at 4 years	Under 30+	43	50	51
	Under 20–30	43	49	52
	Under 10–20	43	50	52
	Under 0–10	40	47	48
	Over 0–5	39	45	49
	Over 5–10	38	42	47
	Over 10–20	37	43	44
	Over 20+	33	39	42
Retained at 6 years	Under 30+	30	38	38
	Under 20–30	29	37	38
	Under 10–20	29	38	37
	Under 0–10	27	36	34
	Over 0–5	26	33	33
	Over 5–10	25	31	32
	Over 10–20	25	31	30
	Over 20+	21	27	27
Promoted to E-5 at YOS6	Under 30+	22	27	29
	Under 20–30	21	25	29
	Under 10–20	20	26	28
	Under 0–10	18	23	24
	Over 0–5	16	21	22
	Over 5–10	15	18	19
	Over 10–20	13	16	16
	Over 20+	9	12	13
Promoted to E-5 at YOS4	Under 30+	14	17	19
	Under 20–30	14	17	19
	Under 10–20	14	17	20
	Under 0–10	12	14	17
	Over 0–5	11	13	16
	Over 5–10	9	12	14
	Over 10–20	8	10	10
	Over 20+	5	7	9

Table 5.7
Predicted Outcomes, by Weight Category and Race/Ethnicity, Navy

Outcome	Weight Relative to Service Standard (lbs)	Percent		
		White (1)	Black (2)	Hispanic (3)
Retained at 3 months	Under 30+	85	88	87
	Under 20–30	85	87	87
	Under 10–20	85	87	87
	Under 0–10	85	87	86
	Over 0–5	84	87	87
	Over 5–10	84	84	87
	Over 10–20	84	86	87
	Over 20+	83	87	89
Retained at 4 years	Under 30+	45	49	49
	Under 20–30	46	49	50
	Under 10–20	46	49	49
	Under 0–10	46	47	49
	Over 0–5	46	49	51
	Over 5–10	45	48	48
	Over 10–20	47	47	49
	Over 20+	44	48	52
Retained at 6 years	Under 30+	26	32	29
	Under 20–30	27	32	28
	Under 10–20	27	31	27
	Under 0–10	28	29	30
	Over 0–5	29	32	31
	Over 5–10	27	30	28
	Over 10–20	29	30	29
	Over 20+	26	31	30
Promoted to E-5 at YOS6	Under 30+	22	20	22
	Under 20–30	22	20	23
	Under 10–20	22	19	22
	Under 0–10	22	18	24
	Over 0–5	22	19	23
	Over 5–10	20	18	22
	Over 10–20	22	19	24
	Over 20+	19	18	19
Promoted to E-5 at YOS4	Under 30+	15	12	14
	Under 20–30	15	12	15
	Under 10–20	15	12	14
	Under 0–10	15	10	15
	Over 0–5	15	9	14
	Over 5–10	14	11	13
	Over 10–20	15	11	13
	Over 20+	14	9	12

Table 5.8
Predicted Outcomes, by Weight Category and Race/Ethnicity, Marine Corps

Outcome	Weight Relative to Service Standard (lbs)	Percent		
		White (1)	Black (2)	Hispanic (3)
Retained at 3 months	Under 30+	88	90	91
	Under 20–30	88	89	90
	Under 10–20	87	88	91
	Under 0–10	87	90	88
	Over 0–5	85	85	89
	Over 5–10	85	85	87
	Over 10–20	85	88	89
	Over 20+	85	83	89
Retained at 4 years	Under 30+	63	65	72
	Under 20–30	62	65	71
	Under 10–20	61	64	71
	Under 0–10	60	64	68
	Over 0–5	57	61	68
	Over 5–10	57	57	66
	Over 10–20	56	59	68
	Over 20+	50	60	69
Retained at 6 years	Under 30+	15	22	22
	Under 20–30	14	20	22
	Under 10–20	13	19	20
	Under 0–10	11	16	18
	Over 0–5	11	17	14
	Over 5–10	9	15	16
	Over 10–20	9	17	14
	Over 20+	6	12	17
Promoted to E-5 at YOS6	Under 30+	20	23	27
	Under 20–30	18	20	25
	Under 10–20	17	20	23
	Under 0–10	14	16	19
	Over 0–5	11	18	15
	Over 5–10	9	14	15
	Over 10–20	9	14	14
	Over 20+	3	13	15
Promoted to E-5 at YOS4	Under 30+	12	11	16
	Under 20–30	12	11	14
	Under 10–20	11	8	13
	Under 0–10	9	9	11
	Over 0–5	7	8	9
	Over 5–10	6	4	8
	Over 10–20	4	7	8
	Over 20+	2	7	5

Table 5.9
Predicted Outcomes, by Weight Category and Race/Ethnicity, Air Force

Outcome	Weight Relative to Service Standard (lbs)	Percent		
		White (1)	Black (2)	Hispanic (3)
Retained at 3 months	Under 30+	94	96	95
	Under 20–30	94	96	97
	Under 10–20	94	95	95
	Under 0–10	94	95	95
	Over 0–5	93	95	96
	Over 5–10	93	94	92
	Over 10–20	94	95	92
	Over 20+	94	93	101
Retained at 4 years	Under 30+	79	81	82
	Under 20–30	79	82	83
	Under 10–20	79	81	83
	Under 0–10	78	81	81
	Over 0–5	76	81	79
	Over 5–10	77	79	76
	Over 10–20	76	84	83
	Over 20+	77	81	82
Retained at 6 years	Under 30+	60	64	63
	Under 20–30	60	64	63
	Under 10–20	59	65	64
	Under 0–10	57	62	60
	Over 0–5	56	60	57
	Over 5–10	55	58	63
	Over 10–20	53	62	63
	Over 20+	54	58	72
Promoted to E-5 at YOS6	Under 30+	36	33	32
	Under 20–30	35	34	35
	Under 10–20	34	33	31
	Under 0–10	32	30	30
	Over 0–5	31	30	28
	Over 5–10	29	28	32
	Over 10–20	32	31	23
	Over 20+	30	30	37
Promoted to E-5 at YOS4	Under 30+	3	2	3
	Under 20–30	3	2	3
	Under 10–20	3	2	2
	Under 0–10	3	2	3
	Over 0–5	3	2	3
	Over 5–10	2	3	3
	Over 10–20	3	2	3
	Over 20+	4	2	2

within five pounds of the standard. The Marine Corps has the most lenient weight standard but requires a strength test for individuals who receive a waiver. It is possible that the Marine Corps' lenient policy toward weight, coupled with the strength requirement, has allowed that service to attract recruits who are well suited to it and therefore are less likely to leave. Interestingly, the four-year Marine Corps retention rate of Hispanic recruits who are more than 20 pounds overweight is not much different than the retention rates of those who are within five pounds of the standard. Apparently, the addition of the strength test weeds out those who are unlikely to complete service.

Summary

This chapter's analysis of career outcomes showed that Hispanics almost always have higher retention rates and faster promotion speeds than do their white counterparts. (The only exception is promotion in the Navy.) The analysis also confirmed past studies that have found that those who are higher quality or who meet or exceed enlistment standards are more likely to stay in service and to be promoted. Furthermore, these race/ethnicity effects were robust to the inclusion of other controls.

Beyond simple race/ethnicity comparisons, this chapter focused on three entry standards (AFQT, education, and weight) because, as shown in previous chapters, these are the three major disqualifying characteristics for Hispanic youth. We find that the effects of these standards on outcomes are relatively small compared with the effects of race and ethnicity. In terms of retention and promotion rates, lower-quality Hispanic recruits often compare well to their higher-quality white counterparts. Their lower AFQT and education and their greater weight are not associated with a pattern of lower retention or generally reduced promotion speed. These results suggest that these Hispanic recruits are better matched to the military or have better opportunities in the military than in the civilian sector.

The results for black recruits are roughly the same as for Hispanic recruits. Retention rates are higher for black recruits in the Army and Air Force and promotion speeds are faster in the Army than for similar white recruits. Similarly, we find that lower-quality black recruits have higher retention rates than similar white recruits who are higher quality or who meet or exceed entry standards. Thus, like Hispanic recruits, black recruits seem to be better suited to the military or to have better opportunities in the military than in the civilian sector. One implication of these results is that lowering AFQT, education, and weight standards while targeting the enlistment of Hispanics would have little effect on retention and promotion rates. But lowering the standards without targeting would adversely affect retention and promotion.

CHAPTER SIX

Increasing Hispanic Enlistments: Evaluating Education and Career Choices

Increasing the supply of Hispanics in the military would require (1) increasing the pool of Hispanics eligible for military enlistment, (2) recruiting more eligible Hispanics, or (3) relaxing current standards so that a larger percent of Hispanic youth qualifies for enlistment.

With regard to the first approach, DoD could actively seek to increase eligibility given current enlistment standards by encouraging greater rates of high school graduation and educational achievement, perhaps by supporting educational and community service programs or by directly coaching and training potential recruits to pass existing standards. The second approach entails increasing efforts to recruit Hispanics who qualify under existing standards by increasing both general and targeted recruiting resources and enhancing military compensation and benefits. The third approach, relaxing standards, was discussed in Chapter Five, which examined the military careers of less qualified recruits.

To help DoD begin to evaluate the effectiveness of the first two approaches to increasing Hispanic enlistments, this chapter uses nationally representative samples of youth drawn primarily from Census data and the 1997 cohort of the National Longitudinal Survey of Youth (NLSY97) to describe the educational and employment choices that Hispanics and other youth make as they transition between adolescence and young adulthood. This descriptive analysis highlights how the Hispanic youth population differs from other racial and ethnic groups in ways that are relevant to military recruiting. However, we do not explicitly evaluate any particular recruiting strategy.

We divide the youth population into three groups: (1) high school dropouts and youth scoring in the 29th percentile or below of the national distribution of scores on the AFQT (Categories IV and V), (2) high school graduates with scores between the 30th and 63rd percentiles of the AFQT distribution (Categories IIIA and IIIB), and (3) high school graduates with scores above the 63rd percentile of the AFQT distribution (Categories I and II). The first group of youth is generally unqualified for military service given current standards (only a small number of Category IV youth are accepted for military service). The bulk of military enlistments come from the second group of youth; this is especially true of black and Hispanic enlistments. The third group of youth is the most desirable and most difficult to recruit.

As noted above, our analyses are based largely on data drawn from the NLSY97, a nationally representative sample of youth ages 12–16 in 1996 surveyed each year between 1997 and

2003.[1] The NLSY97 data allow us to follow a single cohort of youth as they age from adolescence to young adulthood and to describe the educational and employment choices they make along the way. By the 2003 survey wave, 98 percent of the NLSY97 sample was between ages 19 and 23. We define our population of high school graduates as those individuals who received a high school diploma (traditional or alternative) by the 2003 survey date. Conversely, high school dropouts are those youth who had attained at least 19 years of age, but had not reported receiving a high school diploma by their last interview date. The ASVAB was administered to about 79 percent of the NLSY97 sample in 1997 and 1998. We computed age-adjusted nationally normed AFQT scores from the raw ASVAB scores as described in Asch and Loughran (2005).

Table 6.1 shows the weighted percent of male NLSY97 youth falling into each of the three groups by race and ethnicity and the actual number of youth in each group surveyed at least once between 1997 and 2003. Although females make up a growing percentage of military accessions, accessions are still predominantly male (80 percent of recruits according to the 2000 MEPS). In the interest of brevity and clarity, therefore, we limited our analysis to male youth. We further limited our sample to youth interviewed in 2003 with a valid AFQT score and nonmissing information on education. These sample restrictions reduced our sample from 4,599 to 3,100 males.[2] The table shows that a much higher percentage of whites graduate from high school and score in Category II and above than do blacks or Hispanics. Although the NLSY97 oversamples the black and Hispanic population, sample sizes for black and Hispanic high school graduates are small. In the analyses below, we note where small sample sizes prevent us from making statistically meaningful comparison across racial and ethnic groups. This is most likely to happen when we present statistics by age in addition to race and ethnicity.

For some analyses, we also employ the 1979 NLSY, a nationally representative sample of youth ages 14–22 when first surveyed in 1979.[3] Males with valid AFQT scores in the NLSY79

Table 6.1
NLSY97 Analysis Populations, by Race/Ethnicity

Population	White		Black		Hispanic	
	Weighted Percentage	n	Weighted Percentage	n	Weighted Percentage	n
High school dropouts and Category IV–V	28	457	65	514	53	328
High school graduates and Category IIIA–IIIB	30	486	25	183	27	143
High school graduates and Category I–II	43	695	11	74	20	96

SOURCE: NLSY97.

NOTE: Sample restricted to males with nonmissing information on AFQT and education.

---

[1] By design, the NLSY97 oversamples blacks and Hispanics. However, the NLSY97 sample is nationally representative when properly weighted. All tabulations and other analyses below employ sample weights. See the *NLSY97 User's Guide* (2005) for more about the NLSY97.

[2] This sample restriction has virtually no impact on the distribution of the sample by race/ethnicity.

[3] For more information about the NLSY97 and NLSY79, refer to http://www.bls.gov/nls/

were ages 35–43 as of the last survey wave in our data, 2000 (n = 5,212). Consequently, the NLSY79 allowed us to examine some labor market outcomes—wages and earnings—at much later ages than we could using the NLSY97. We also relied on the 2000 MEPS for statistics on the AFQT distribution of military applicants and enlistees.[4]

The next four sections of the chapter describe the educational and employment choices of our three groups of youth: high school dropouts and Category IV–V youth, Category III high school graduates, and Category I–II high school graduates. The final section discusses how these descriptive analyses can inform DoD as it implements policies and strategies to improve recruiting in the Hispanic population.

High School Dropouts and Category IV–V Youth

The U.S. military has long required its recruits to possess a high school diploma or equivalent certificate of high school completion. In fiscal year (FY) 2002, fewer than 1 percent of recruits had not completed high school as of the beginning of their enlistment (Department of Defense, 2004). The reason for maintaining this educational standard is that high school graduates are far less likely to attrite during their first term of enlistment and more likely to reenlist following it than are high school dropouts. This is true even conditional on AFQT (Buddin, 2005). For Hispanic males, this educational standard represents a significant barrier to enlistment. In Chapter Three, we reported in Table 3.2 that only 51 percent of Hispanic males ages 17–21 who were not currently attending high school possessed a high school diploma in 2000, compared with 68 and 84 percent of black and white male youth, respectively.

But, as we showed in Chapter Three, not only lack of a high school diploma limits military eligibility. A large percentage of blacks and Hispanic high school graduates also fail to meet minimum AFQT standards. About 53 percent of black males and 42 percent of Hispanic males in the youth population are Category IV or V, as shown in Table 6.2. DoD accepts only a small percent of Category IV applicants and virtually no Category V applicants.[5] In 2000, for example, 15 percent of all applicants were Category IV or V, but only 1 percent of these applicants eventually enlisted (see Table 6.2).

High school dropouts and those youth who score poorly on the AFQT are also less likely to meet other enlistment standards, such as height/weight standards and moral requirements related to criminal activity and drug use. For example, between 28 and 45 percent of this group of Hispanics fail to meet the services' height/weight standards (authors' computations). By comparison, between 22 and 38 percent of male Hispanic high school graduates scoring in Category IIIB and above fail the services' height/weight standards. According to the 2003

---

[4]   See Chapter Five for a description of this data source.

[5]   By law, the services cannot accept Category V youth, and no more than 20 percent of accessions can be Category IV. Category IV accessions must be high school graduates. DoD guidance states that no more than 4 percent of accessions can be Category IV and the services limit this percentage even further (Department of Defense, 2004).

Table 6.2
Distribution of White, Black, and Hispanic High School Graduates, by AFQT Category (%)

AFQT Category	White			Black			Hispanic			All		
	Population	DoD Applicants	Enlistees	Population	DoD Applicants	Enlistees	Population	DoD Applicants	Enlistees	Population	DoD Applicants	Enlistees
I	12	5	6	2	1	1	3	1	2	10	4	4
II	35	36	40	12	14	19	22	18	23	31	29	34
IIIA	16	25	27	11	19	26	14	22	27	15	23	27
IIIB	17	25	25	22	39	51	19	37	46	18	30	33
IV	15	8	1	35	23	3	33	19	2	19	13	1
V	5	1	0	18	4	0	9	4	0	7	2	0

SOURCES: NLSY97; 2000 MEPS.

NOTES: DoD sample is restricted to male applicants ages 17–23 and male enlistees ages 18–24 in 2000. NLSY97 sample is restricted to males ages 18–24 in 2003 with a high school education. "Population" = overall population. "All" includes all applicants, not just whites, blacks, and Hispanics. Columns may not sum to 100 due to rounding.

Youth Poll, as many as 27 percent of male Hispanic dropouts could be disqualified from military service because of recent drug use and other criminal activity.[6] This compares to about 14 percent of male Hispanic high school graduates.

High school dropouts and Category IV and V youth display poor labor market outcomes as well, and obviously, poorer educational outcomes. Not only are employment rates of high school dropouts and Category IV and V youth lower than those of high school graduates, especially among blacks (Figure 6.1), but the wages these youth earn are significantly lower. Even at young ages, the wages of high school dropouts and Category IV and V youth lag behind those of high school graduates. At age 22, for example, the wages of this group of Hispanics are about 20 percent less than the wages of Hispanic high school graduates. As we will see later in Figure 6.15, this difference in wages by AFQT category only grows over time.

Thus, while relaxing high school graduation and AFQT requirements would increase the percentage of Hispanic youth eligible to enlist in the military, doing so could result in an increase in undesirable first-term outcomes, given the relationship between graduation status and first-term outcomes. Perhaps, however, DoD might be able to devise policies that would help raise the educational outcomes and AFQT scores of Hispanic youth. It is clear that such policies would need to overcome deeply rooted family background characteristics that are strongly negatively correlated with high school graduation and AFQT scores (Swail, Cabrera, and Lee, 2004; Cameron and Heckman, 2001). For example, much of the gap in high school graduation rates between Hispanics and blacks appears to be attributable to immigration status. According to the 2000 Census, fewer than half (48 percent) of Hispanic males ages 18–24 were born in the United States, and the high school graduation rate of non–U.S. born Hispanics is very low (37 percent). U.S.-born Hispanic males ages 18–24, on the other hand, are just as likely to graduate from high school as are U.S.-born blacks (63 percent of both groups have graduated from high school). According to Fry (2003), many non-U.S.-born Hispanic high school dropouts never attended school in the United States. Not surprisingly, Hispanic dropouts tend to have poor English language skills, which could also help account for their failure to complete high school in the United States (Fry, 2003).

Other family background characteristics, such as family income and the educational attainment of mothers and fathers, account for about half the gap in high school graduation rates and AFQT scores between whites on the one hand and blacks and Hispanics on the other. Table 6.3 reports the results of a linear regression of high school graduation and AFQT score on race/ethnicity, age, immigration status, whether a child lived with both biological parents in 1997, mother's education, father's education, family income in 1997 as a percentage of the federal poverty line, and census region of residence in 1997 (the dependent variable equals 1 if the respondent graduated from high school and scored above Category IV on the AFQT). Column (1) of the table reports regression results that omit any family background characteristics. The coefficient on blacks of −0.39 and the coefficient on Hispanics of −0.31 indicate that blacks and Hispanics are 39 and 31 percentage points less likely than whites to graduate from high school and score above Category IV on the AFQT. Column (2) controls for the family background variables listed above. The coefficient on blacks falls to −0.26 and the coefficient

---

[6] These computations are based on analyses reported in Chapter Three.

Figure 6.1
Percentage Currently Employed, by Race/Ethnicity, Age, and Education/AFQT

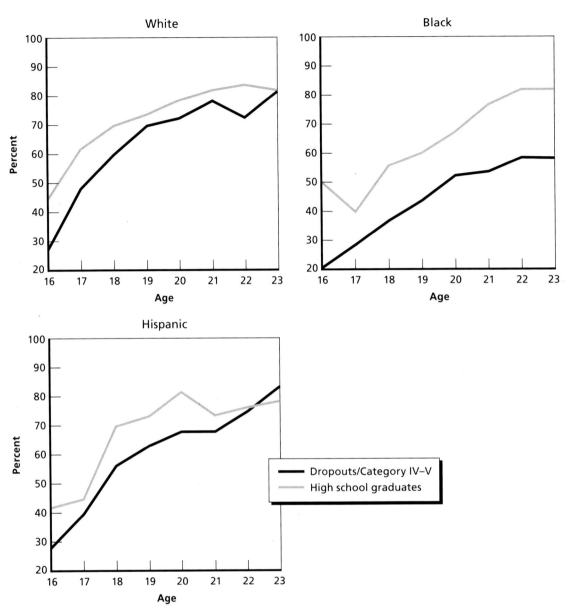

NOTE: Sample restricted to males.
RAND MG773-6.1

on Hispanics falls to –0.14, indicating that differences in family background have a strong impact on differences in the high school graduation rate and AFQT scores of whites, blacks, and Hispanics. Thus, it is apparent that family background characteristics are an important determinant of high school graduation rates and AFQT scores. DoD policies that aim to help blacks and Hispanics graduate from high school at rates comparable to whites or to score as well on the AFQT must counter these powerful forces.

Table 6.3
The Effect of Family Background on High School Graduation and AFQT Scores

	(1)	(2)
Black	−0.39	−0.26
	(0.02)**	(0.02)**
Hispanic	−0.31	−0.14
	(0.02)**	(0.02)**
Other race/ethnicity	−0.01	0.03
	(0.04)	(0.04)
Age	0.02	0.02
	(0.01)**	(0.01)**
Constant	0.36	−0.05
	(0.13)**	(0.13)
s		
R^2	0.13	0.25
n	3,062	3,062

SOURCE: NLSY97.

NOTES: Dependent variable = 1 if graduated high school by 2003 interview and scored above Category IV on the AFQT. White is the excluded race/ethnicity category. Column (2) controls for the family background variables listed in the text.

** Significant at 1 percent.

On a more positive note, the supply of Hispanic high school graduates should continue to increase in the years to come, both because the sheer number of Hispanics in the population is growing and because, just as with earlier waves of immigration, second- and third-generation Hispanics (i.e., the children and grandchildren of first-generation Hispanic immigrants) are much more likely to graduate from high school than are first-generation Hispanics (Smith, 1993). So, while high school graduation rates are likely to prevent a large number of recent Hispanic immigrants from joining the military, this obstacle should apply to a falling percentage of the Hispanic population in coming decades. Presumably, the same will be true with respect to AFQT scores.

High School Graduates, Category IIIA–IIIB

The majority of the enlisted population is Category III. This is especially true of the black and Hispanic enlisted populations. Table 6.4 reports the percentage of the youth population, military applicants, and military enlistees scoring in Categories I–IIIB. We have restricted the population in all cases to those scoring above Category IV, since Category V youth are ineligible to enlist, and the services accept only a small percentage of Category IV applicants. In 2000, 1, 3, and 2 percent of white, black, and Hispanic accessions, respectively, were Category IV.

Table 6.4
Distribution of White, Black, and Hispanic High School Graduates, by AFQT Category: AFQT ≥ Category IIIB

AFQT Category	White			Black			Hispanic			All		
	Population	DoD Applicants	Enlistees	Population	DoD Applicants	Enlistees	Population	DoD Applicants	Enlistees	Population	DoD Applicants	Enlistees
I	15	5	6	3	1	1	5	2	2	14	4	4
II	44	39	41	26	19	20	37	23	23	42	34	35
IIIA	20	27	28	24	26	27	24	28	28	21	27	28
IIIB	21	28	26	47	54	53	32	48	47	24	35	33

SOURCES: NLSY97; 2000 MEPS.

NOTES: DoD sample is restricted to male applicants ages 17–23 and male enlistees ages 18–24 in 2000, Category IIIB or higher. NLSY97 sample is restricted to males ages 18–24 in 2003 with a high school education, Category IIIB or higher. "All" includes all applicants, not just whites, blacks, and Hispanics. Columns may not sum to 100 due to rounding.

The AFQT distribution of the overall youth population, applicants, and enlistees varies dramatically by race. About 26 percent of the sample of enlisted whites, for example, are Category IIIB, compared with 53 and 47 percent of blacks and Hispanics, respectively. Combined with the fact that black and Hispanic DoD applicants also score much lower than whites on the AFQT (e.g., 28, 54, and 48 percent of white, black, and Hispanic applicants are Category IIIB), these statistics suggests that in order for the military to maintain a force of the desired size with a reasonably balanced proportion of whites, blacks, and Hispanics, it must draw a relatively large percentage of blacks and Hispanics from the lower half of the AFQT distribution.

It is also clear from Table 6.4 that the propensity of male high school graduates to apply for military service varies considerably by AFQT. Overall, about 14 percent of the youth population Category IIIB and above is Category I, but only 4 percent of DoD applicants are Category I. Conversely, 24 percent of the youth population Category IIIB and above is Category IIIB compared with 35 percent of DoD applicants. Thus, in general, the propensity to apply to the military increases as AFQT falls (see Table 6.5).

One explanation for this pattern could be that individuals with lower AFQT scores find the military to be a more attractive employment opportunity than do individuals with higher scores, since the military offers a comparatively more attractive wage for lower-AFQT individuals. This is evident in Figure 6.2, which plots annual civilian earnings of full-time[7] male high school graduates ages 30–34 surveyed by the NLSY79[8] by AFQT category and RMC for service members with between 12 and 16 years of experience (who, roughly speaking, should be between 30 and 34 years of age).[9] We focus on this age group, since these individuals are likely to have settled in to long-term careers and their annual earnings offer a reasonable measure of long-term earnings potential. The figure shows the 25th, 50th, and 75th percentiles of the civilian earnings distribution by AFQT and three levels of regular military

Table 6.5
Propensity to Apply Relative to Category IIIB Youth, by Race/Ethnicity and AFQT

AFQT Category	White	Black	Hispanic	All
I	0.25	0.29	0.27	0.20
II	0.66	0.64	0.41	0.56
IIIA	1.01	0.94	0.78	0.88
IIIB	1	1	1	1

SOURCES: NLSY97; 2000 MEPS.

NOTES: Figures derived from Table 5.3 by dividing applicant percentage by population percentage and then dividing that figure by the corresponding figure for Category IIIB youth within race/ethnic groups. Thus, the propensity to enlist is relative to Category IIIB youth of the same racial/ethnic background. "All" includes all applicants, not just whites, blacks, and Hispanics.

---

[7]   Full-time workers are defined as those working at least 1,664 hours per year.

[8]   These individuals were all surveyed after 1996.

[9]   We show three levels of RMC corresponding to individuals who excelled, are about average, or fall below expected rank at age 30–34 (corresponding to 12 to 16 YOS).

Figure 6.2
Annual Civilian and Military Earnings, by AFQT ($000)

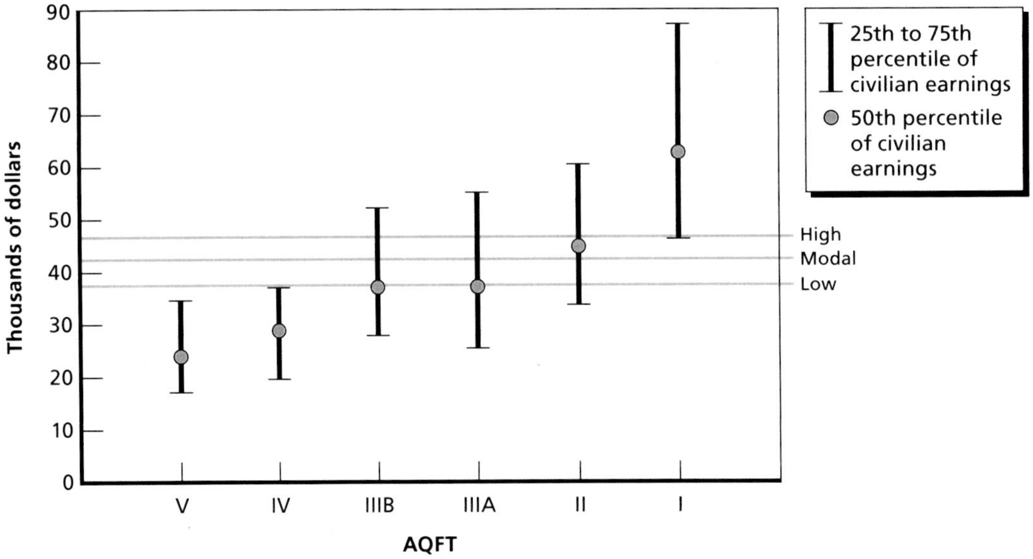

SOURCES: NLSY79; DoD (2001).
NOTES: Sample restricted to males ages 30–34 with a high school diploma or more, working at least 1,664 hours per year. Military earnings are denoted by the horizontal lines. "High" RMC is for an E-7, YOS 14 in 2001. "Modal" RMC is for an E-6, YOS 14 in 2001. "Low" RMC is for an E-5, YOS 14 in 2001. All figures are in 2001 dollars.
RAND *MG773-6.2*

compensation (RMC) corresponding to individuals who excelled, are about average, or fall below expected rank at 14 years of service.[10] All figures are reported in 2001 dollars.

The figure shows that median civilian earnings generally increases with AFQT. Median earnings for a Category V individual are $24,400 compared with $62,900 for a Category I individual. The figure also shows that modal RMC (indicated by the middle horizontal line) compares favorably to median civilian earnings for all but Category I individuals.[11] Of course, civilian earnings can easily exceed modal RMC, and this is more likely to happen among high AFQT individuals (the 25th-75th percentile of the civilian earnings distribution is denoted by the vertical lines).

We also noted earlier that a much higher percentage of black and Hispanic enlistees are Category IIIB than are white enlistees; 26, 53, and 47 percent of white, black, and Hispanic enlistees are Category IIIB. Roughly equal percentages of white, black, and Hispanic enlistees are Category IIIA (about 28 percent). In this section, we compare the educational and labor market outcomes of Category IIIA and IIIB youth as they transition from adolescence to young adulthood to see whether there are differences between these two AFQT groups by race/ethnicity.

First, we examine transitions into two- and four-year college by race/ethnicity and AFQT (Figures 6.3–6.4). Looking at whites first, we see that two-year college enrollment rates peak

---

[10] RMC is derived from DoD (2001). High" RMC is for an E-7, YOS 14 in 2001. "Modal" RMC is for an E-6, YOS 14 in 2001. "Low" RMC is for an E-5, YOS 14 in 2001.

[11] The comparison of civilian earnings to RMC is likely to be even more favorable to the military now. In 2002, DoD significantly increased basic pay and reduced out-of-pocket housing costs, thereby increasing RMC considerably.

Figure 6.3
Percentage Enrolled in a Two-Year College Program, by Race/Ethnicity and AFQT

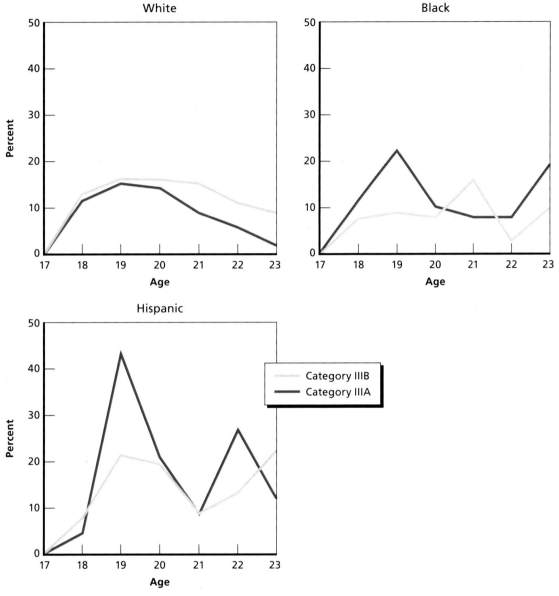

NOTE: Sample restricted to male high school graduates, Category IIIA-IIIB.
RAND *MG773-6.3*

at about 16 percent at age 19 for both Category IIIA and IIIB youth. Four-year enrollment rates peak at 32 percent for Category IIIA whites at age 21 and 15 percent at age 19 for Category IIIB whites. Two-year college enrollment rates are slightly higher for Category IIIB than for Category IIIA, but four-year college enrollment rates are considerably higher for Category IIIA than for Category IIIB. At the oldest age, we observe these individuals in the data (an average of 21), 45 and 57 percent of Category IIIA and IIIB white youth, respectively,

Figure 6.4
Percentage Enrolled in a Four-Year College Program, by Race/Ethnicity and AFQT

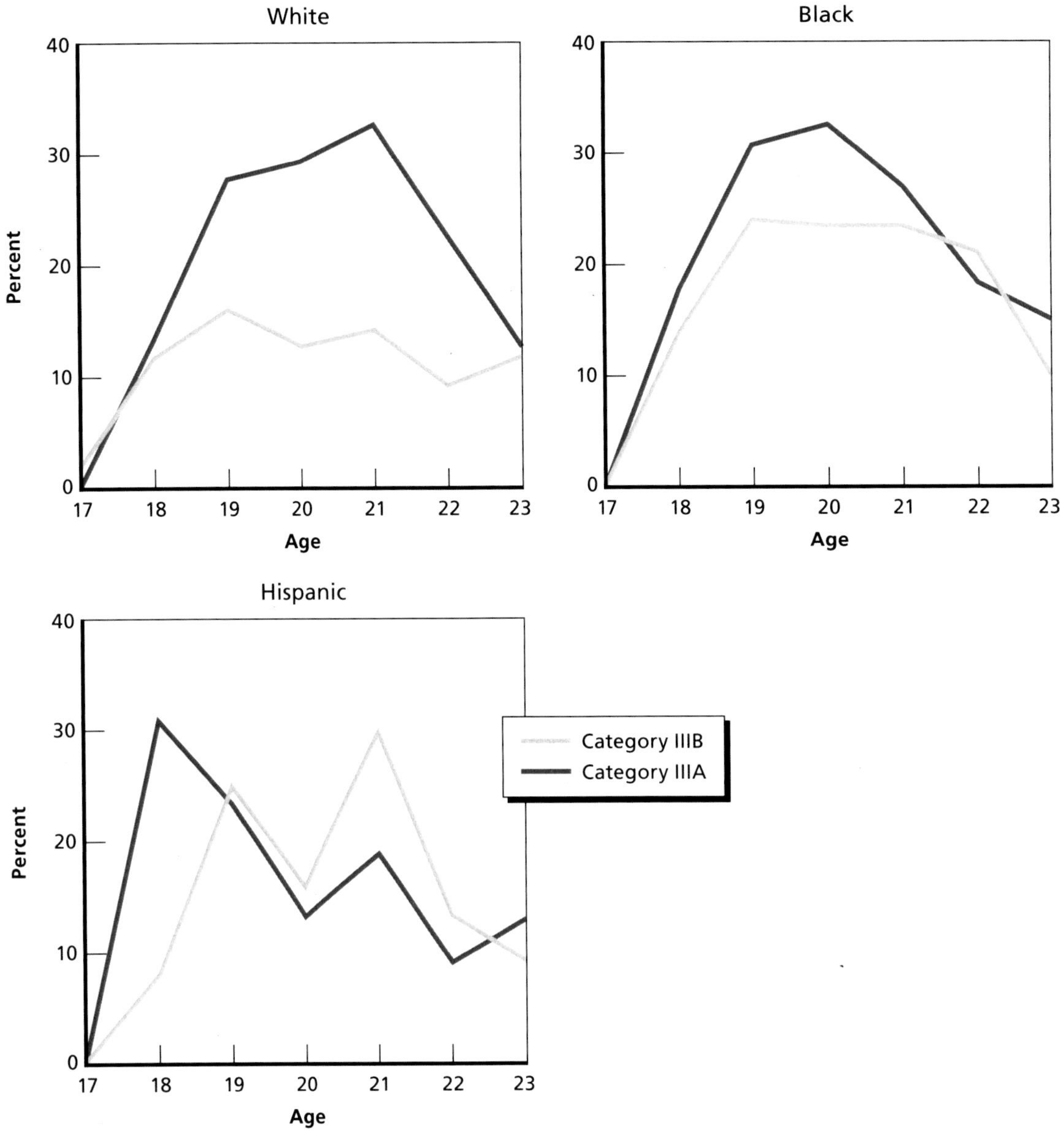

NOTE: Sample restricted to male high school graduates, Category IIIA–IIIB.
RAND MG773-6.4

(Table 6.6). The same enrollment patterns hold for blacks, although overall enrollment rates are slightly higher (at their oldest age in the survey, 56 and 60 percent of black Category IIIA and IIIB youth, respectively, had ever enrolled in college).

The enrollment patterns are somewhat different for Hispanics. First, two-year college enrollment rates are considerably higher for Hispanics than for either blacks or whites, but four-year enrollment rates are lower.[12] Second, unlike for whites and blacks, Cate-

---

[12] Swail, Cabrera, and Lee (2004) report a similar finding in data drawn from the National Educational Longitudinal Study (NELS) as do Ganderton and Santos (1995) using data drawn from the High School and Beyond Survey (HSB).

Table 6.6
Percentage Ever Enrolled in College,
by Race/Ethnicity and AFQT

Category	White	Black	Hispanic
Two-year			
IIIB	29	29	40
IIIA	28	25	50
Four-year			
IIIB	18	31	26
IIIA	35	41	27
Either			
IIIB	45	56	62
IIIA	57	60	63

SOURCE: NLSY97.

NOTE: Sample restricted to Category IIIA–IIIB male
high school graduates in their last survey year.

gory IIIA Hispanics have higher two-year enrollment rates than Category IIIB Hispanics. Four-year enrollment rates are similar for Category IIIA and IIIB Hispanics.

While a considerable percentage of Category IIIA and IIIB youth enroll in college, comparatively few actually complete a degree by age 22 or 23. Only 1 and 8 percent of Category III youth ever enrolled in a two-year or four-year college program, respectively, had completed an Associate's or Bachelor's degree by age 22 or 23. Asch and Loughran (2005) report low completion college rates in the NLSY79 where they can observe respondents at older ages. In that sample, 23 percent of Category III males enrolled in a two-year college program between ages 17 and 21 had completed an Associate's degree six years after initial enrollment; 36 percent of Category III males enrolled in a four-year college between those ages had completed a Bachelor's degree within six years of initial enrollment. Hispanic and black males are less likely than whites to complete a two- or four-year degree within six years of first enrolling.

Although few NLSY97 respondents had completed college by the last survey wave, a much larger number had received some type of professional certification or vocational license as a result of a training program. Table 6.7 lists the percentage of Category IIIA and IIIB youth by race/ethnicity who had received a formal training certification or vocational license by their

Table 6.7
Percentage Receiving Training Certification or Vocational
License, by Race/Ethnicity and AFQT

AFQT	White	Black	Hispanic
IIIB	16	25	23
IIIA	20	13	31
IIIB and IIIA	18	21	26

SOURCE: NLSY97.

NOTE: Sample restricted to Category IIIA–IIIB male high school
graduates in their last survey year.

last survey wave. Overall, Hispanics are the most likely to have received such a certification (26 percent compared with 18 and 21 percent of whites and blacks, respectively). For whites and Hispanics, Category IIIA youth are more likely to have received a training certificate than Category IIIB youth; for blacks, the reverse is true.

A large percentage of Category IIIA and IIIB youth have had some sort of employment experience by the time they graduate from high school. In our sample, about 90 percent of both Category IIIA and IIIB youth were ever employed by age 17. Current employment, graphed by age in Figure 6.5, trends upward with age for all three race/ethnic groups. Overall, 43 percent of youth were employed at age 16. This percentage rises to 80 percent by age 23. Controlling for AFQT, the employment rate of blacks is about 14 percentage points less than that of whites; the employment rate of Hispanics is about 8 percentage points less than that of whites.

Annual hours of work also trend upward with age, again with little difference between Category IIIA and IIIB individuals or across the three race/ethnic groups (Figure 6.6). By age 22 or 23, Category III individuals on average are working 1,445 hours/year, which is considerable, given that 24 percent of these individuals are currently enrolled in college. One notable aspect of Figure 6.6 is the divergence in annual hours worked between Category IIIA and IIIB black youth after age 20. At age 22 and 23, annual hours of work average 1,504 for Category IIIA blacks and 1,038 for Category IIIB blacks.

Although employment increases with age, school enrollment falls, and the net result of these two trends is an increase in idleness with age, where *being idle* refers to being neither enrolled in college nor currently employed (Figure 6.7). By age 22 or 23, about 13 percent of Category III youth are idle. The overall percentage of youth idle at any given age does not vary appreciably between whites and Hispanics or by AFQT in these two groups. For blacks, however, we observe both relatively low levels of idleness after age 20 for Category IIIA youth and relatively high levels of idleness for Category IIIB youth.

Working while attending college is common among Category III youth ages 16–23. About 69, 58, and 70 percent of Category III whites, blacks, and Hispanics work while attending college. Annuals hours of work average 1,008, 753, and 1,051 for whites, blacks, and Hispanics attending college, respectively.

Figures 6.8 and 6.9 plot tenure on the respondent's main job and the number of different employers a respondent has ever had by age, race/ethnicity and AFQT. Figure 6.8 suggests that these youth tend to work for many employers in the early stages of their labor market careers. By age 22 or 23, these youth on average have worked for seven different employers. In the case of black youth, there is some divergence in number of employers ever had between Category IIIA and IIIB youth after age 20. Despite having worked for many employers over their early careers, these youth have on average worked on what they consider to be their main job for close to two years by age 22 or 23.[13] Among blacks we see that Category IIIA youth have higher tenure on their main job than Category IIIB youth. For whites and Hispanics there appears to be little difference in tenure between Category IIIA and IIIB youth.

---

[13] The main job is the job at which the respondent works the most hours.

Figure 6.5
Percentage Currently Employed, by Race/Ethnicity and AFQT

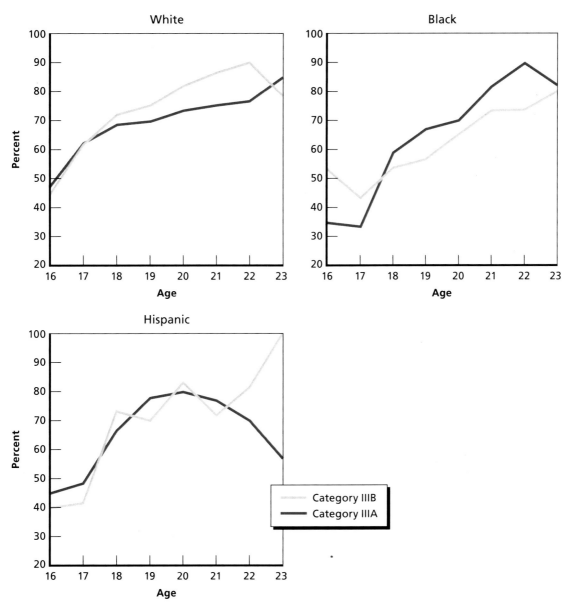

NOTE: Sample restricted to male high school graduates, Category IIIA–IIIB.
RAND *MG773-6.5*

Figure 6.6
Annual Hours of Work, by Race/Ethnicity and AFQT

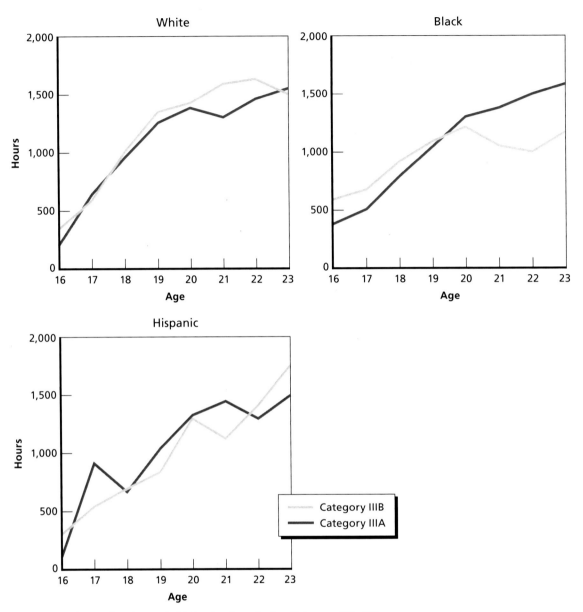

NOTE: Sample restricted to male high school graduates, Category IIIB–IIIA.

Figure 6.7
Percentage Idle, by Race/Ethnicity and AFQT

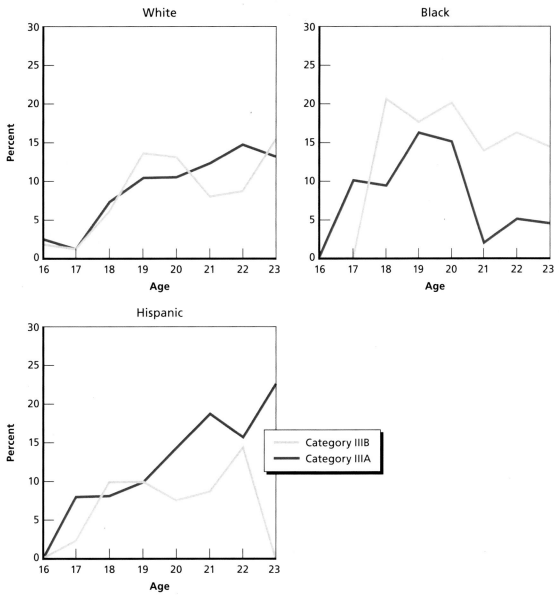

NOTES: "Idle" is defined as being neither enrolled in college nor currently employed. Sample restricted to male high school graduates, Category IIIA–IIIB.

RAND *MG773-6.7*

Figure 6.8
Number of Employers Worked For Since Age 16, by Race/Ethnicity and AFQT

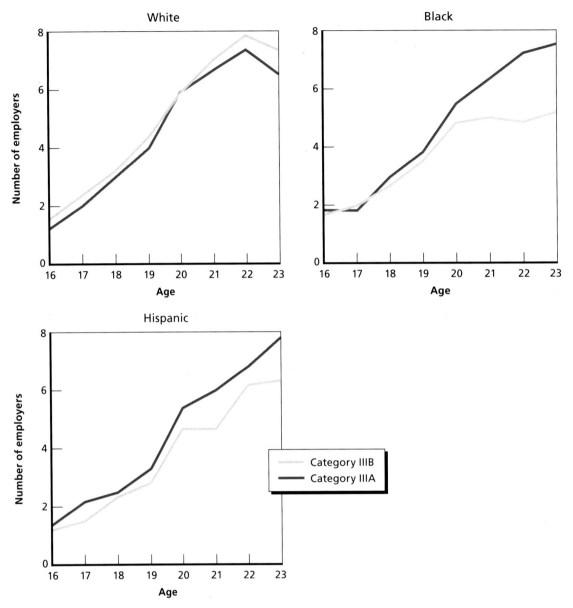

NOTES: Sample restricted to male high school graduates, Category IIIA–IIIB.
RAND *MG773-6.8*

Figure 6.9
Job Tenure, by Race/Ethnicity and AFQT (Weeks)

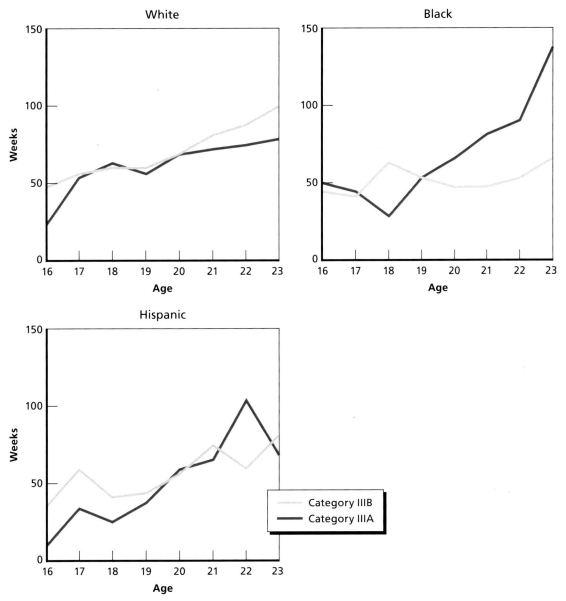

NOTES: Tenure is for the respondent's "main" job. Sample restricted to male high school graduates, Category IIIA–IIIB.

RAND *MG773-6.9*

Figure 6.10
Median Hourly Wages, by Age, Race/Ethnicity, and AFQT, NLSY97

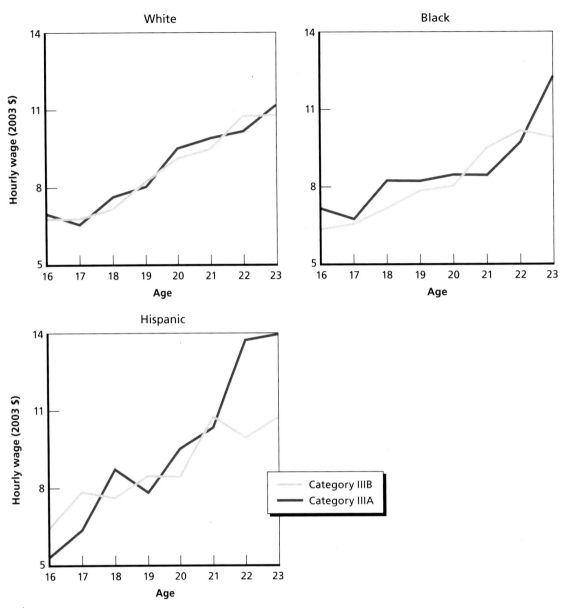

NOTE: Sample restricted to male high school graduates, Category IIIA–IIIB.
RAND *MG773-6.10*

Real median hourly wages trend steadily upward with age for all three race/ethnic groups in Category IIIA and IIIB. This fact is evident in Figure 6.10, plotted from the NLSY97, that tracks wages between ages 16 and 23; and Figure 6.11, plotted from the NLSY79, that tracks wages over a longer period—between ages 17 and 43. In the NLSY97, median real wages increase by about 57 percent between ages 16 and 23. In the NLSY79, median real wages approximately triple between ages 17 and 43. There is no perceptible difference in the median wages of Category IIIA and IIIB youth in the NLSY97. In the NLSY79, the median wages of Category IIIB youth appear to lag slightly behind those of Category IIIA youth beginning in

Figure 6.11
Median Hourly Wages, by Age, Race/Ethnicity, and AFQT, NLSY79

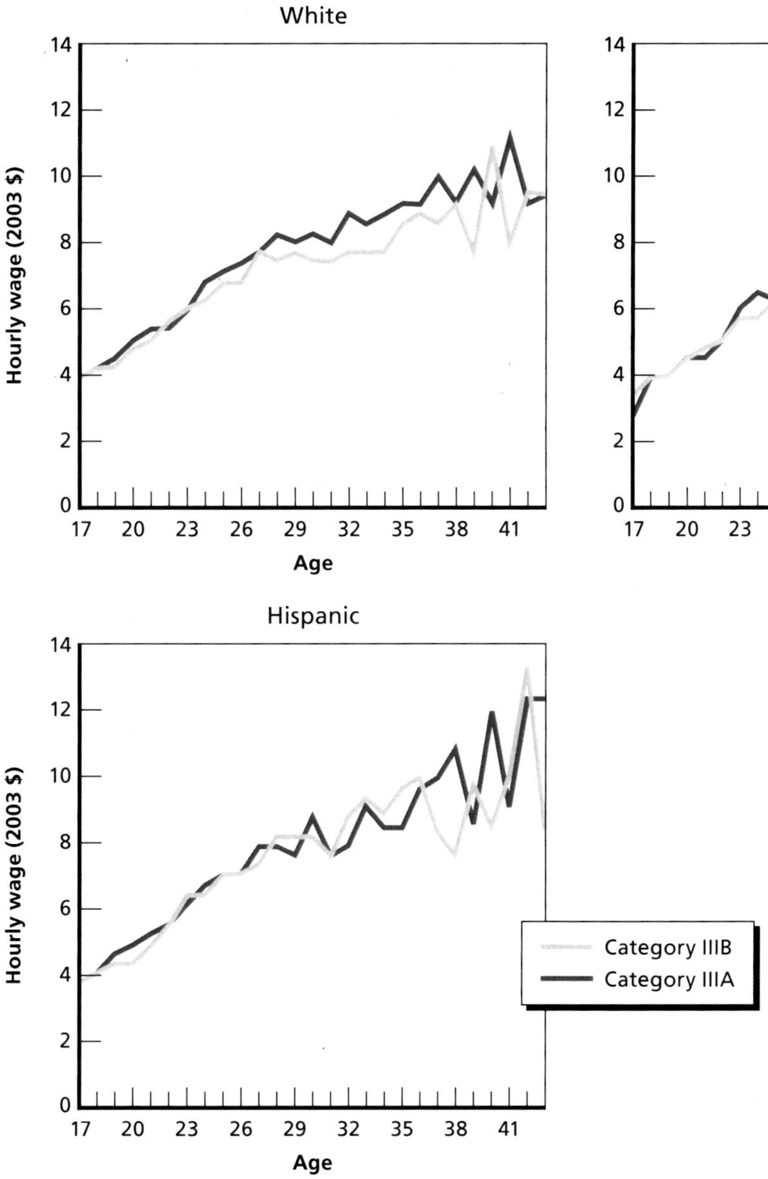

NOTE: Sample restricted to male high school graduates, Category IIIA–IIIB.
RAND *MG773-6.11*

their mid-twenties. This is true for whites and blacks, but not for Hispanics.[14] In neither sample do we detect statistically significant differences in wages between the three racial/ethnic groups once we condition on AFQT.[15]

---

[14] Specifically, we cannot reject the hypothesis that Category IIIA and IIIB wages differ after age 25 in the Hispanic population. We can reject this hypothesis in the white and black populations. The apparent "noisiness" of the wage graphs at later ages is due to small sample sizes.

[15] These same findings hold if we examine wages at the 25th and 75th percentiles rather than at the median.

Finally, we look at marriage and childbearing in this population. Table 6.8 shows that, by their last survey wave in the NLSY97, fewer than 10 percent of Category IIIA and IIIB youth had ever been married across all groups. Marriage rates are somewhat higher for whites and Hispanics than they are for blacks. The percentage of these youth who have a child by that age is also low, about 8 percent of whites, 13 percent of blacks and 14 percent of Hispanics. Among Hispanics, Category IIIB youth are more likely than Category IIIA youth to have ever had a child.

In summary, the early labor market and educational experiences of white, black, and Hispanic Category IIIA and IIIB high school graduates are remarkably similar. A considerable percentage of these youth pursue some sort of postsecondary education, but relatively few actually complete a degree by age 22. A much higher percentage, though, do complete some sort of formal training and receive a training certification or a vocational license. The employment rate averages about 74 percent between ages 18 and 23 in this population, and annual hours of work increase steadily between ages 16 and 23. Median wages also increase steadily between ages 16 and 23—by about 57 percent. These youth hold a large number of different jobs during this period, consistent with the findings of Karoly and Klerman (1994) and Neumark (2002) who examined the same phenomenon using the NLSY79. Neumark (2002) argues that this type of job "churning" is natural as young people seek appropriate employment for themselves. Despite this job churning, tenure on the respondent's main job increases with age, suggesting that these youth are beginning to settle into more stable employment by their early twenties.

With respect to Hispanics in particular, one notable difference between these youth and white and black youth is their comparatively high propensity to attend a two-year college and relatively low propensity to attend a four-year college. Otherwise, Hispanics look remarkably similar to whites in this population, except for their somewhat higher propensity to have fathered a child.

There appears to be little difference between Category IIIA and IIIB individuals, at least in these early years. Overall, Category IIIA individuals are somewhat more likely to attend a four-year college than are Category IIIB individuals, but college completion rates are similar. Category IIIB blacks are somewhat more likely not to be currently employed or enrolled in school than are Category IIIA blacks. Perhaps most salient, median hourly wages are about the same for those Category IIIB and IIIA youth who do work, although Category

Table 6.8
Percentage Ever Having Been Married or Ever Had a Child,
by Race/Ethnicity and AFQT

AFQT	White	Black	Hispanic
Married			
IIIB	9	7	11
IIIA	10	6	11
Ever had a child			
IIIB	7	12	19
IIIA	9	16	8

SOURCE: NLSY97.

NOTE: Sample restricted to Category IIIA–IIIB male high school graduates.

IIIB wages begin to lag slightly behind those of Category IIIA individuals in their mid to late twenties. For Hispanics, though, median wages do not differ between Category IIIA and IIIB youth, even at later ages, as shown in Figure 6.11.

High School Graduates, Category I–II

In 2000, Category I and II high school graduates comprised 47, 14, and 25 percent of all white, black, and Hispanic male applicants, respectively, (Table 6.2). Compared with Category III youth, these individuals are considerably less likely to apply for military service (Table 6.4). Overall, the propensity to apply of Category I and II youth is about 80 and 44 percent less than the propensity to apply of Category IIIB youth (Table 6.5). This fact is consistent with the fact that Category I and II youth are likely to have considerably better labor market opportunities in the civilian sector than they are in the military (see Figure 6.2). In this section, we document the extent to which these superior civilian labor market opportunities are apparent even at early ages. Although there are substantial differences between Category I and II youth, we consider these individuals collectively since there are insufficient numbers of Category I black and Hispanic youth in our data for statistical analysis.

We begin by examining enrollment in two- and four-year colleges. Category I and II youth are about twice as likely to attend a four-year college as are Category III youth, but their likelihood of attending a two-year college is considerably less. Whereas we did not observe substantial differences in enrollment rates across Category III youth by race/ethnicity, such differences are more apparent among Category I and II youth (Figure 6.12). In particular, by their last survey wave, 67 percent of whites had ever enrolled in a four-year college, compared with 63 percent of blacks and 52 percent of Hispanics.[16] On the other hand, Category I and II Hispanics are somewhat more likely to attend a two-year college than are whites and blacks. College completion rates are much higher for Category I and II youth than they are for Category III youth. This is most apparent when we look at the NLSY79. Among youth of that cohort, 80 percent of Category I youth and 63 percent of Category II youth finish a four-year degree within six years of initial enrollment (Asch and Loughran, 2005). The percentage of Category I and II youth receiving some sort of training certificate or vocational license is about the same as it is for Category III youth (about 17 percent by their last wave of the survey).

As with Category III youth, employment rates of Category I and II youth increase steadily with age in our sample (Figure 6.13, first panel). By age 22 or 23, 80 percent of the overall sample is employed. Annual hours employed also increase steadily with age in our sample. By age 22 or 23, Category I and II youth are working an average of 1,316 hours per year, which is slightly less than the number of hours worked by Category III youth (Figure 6.13, second panel). Employment and annual hours of work for black Category I and II youth lag notably behind those for Category I and II white and Hispanic youth.

---

[16] Some of this difference is attributable to the fact that whites are more likely to be Category I than are blacks or Hispanics. About 62 percent of Category II whites have ever enrolled in a four-year college in our sample.

Figure 6.12
Two- and Four-Year College Enrollment, by Age and Race/Ethnicity

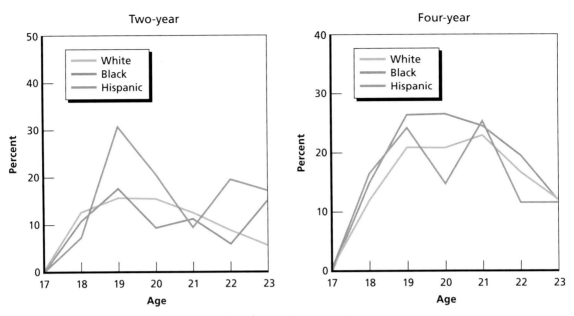

NOTE: Sample restricted to male high school graduates, Category I–II.
RAND *MG773-6.12*

Figure 6.13
Percentage Currently Employed and Annual Hours Employed, by Age and Race/Ethnicity

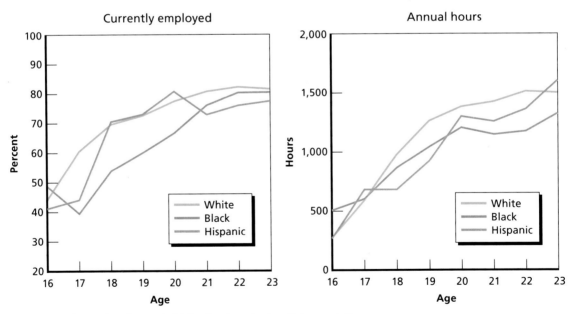

NOTE: Sample restricted to male high school graduates, Category I–II.
RAND *MG773-6.13*

The relatively low level of annual hours employed for Category I and II blacks is also evident in tenure on the main job (Figure 6.14, first panel). In their main job, blacks have lower tenure than either whites or Hispanics. As with Category III youth, Category I and II youth hold a number of jobs in their early careers (Figure 6.14, second panel). By age 22, the mean number of jobs ever held is six. Category I and II whites appear to have held slightly more jobs than either blacks or Hispanics, but the difference is small.

At ages 22 or 23, the median hourly wages of Category I and II youth are roughly comparable to the median hourly wages of Category III youth (as shown in Figure 6.15). As these youth age, however, it is likely that their wages will surpass those of Category III youth. We make this inference in part based upon Figure 6.15, which is plotted from the NLSY79 and shows that age-wage profiles generally steepen as AFQT increases.

Figure 6.16 shows that there is little difference in the median hourly wages of Category I and II youth in the NLSY97 by race/ethnicity. We also see this in the NSLY79 at later ages (Figure 6.17). In Table 6.9, we present regressions that isolate the variation in log wages attributable to race/ethnicity and other factors. Our sample consists of males ages 25 and older with at least a high school diploma and nonmissing data on AFQT and hourly wages.[17] In column (1), we see that, conditional on educational attainment, black males earn hourly wages that

Figure 6.14
Average Tenure on the Main Job (Weeks) and Number of Jobs Held Since Age 16, by Age and Race/Ethnicity

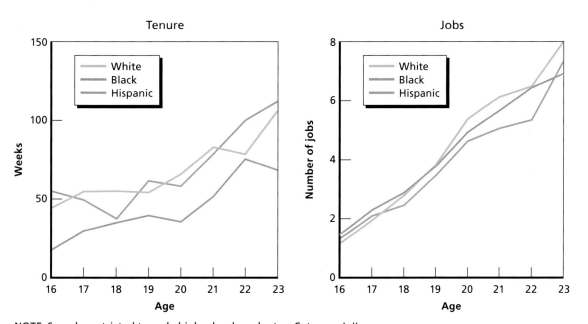

NOTE: Sample restricted to male high school graduates, Category I–II.
RAND *MG773-6.14*

---

[17] Our data include multiple observations per person. Standard errors are adjusted accordingly.

Figure 6.15
Median Hourly Wage, by Age and AFQT: NLSY79

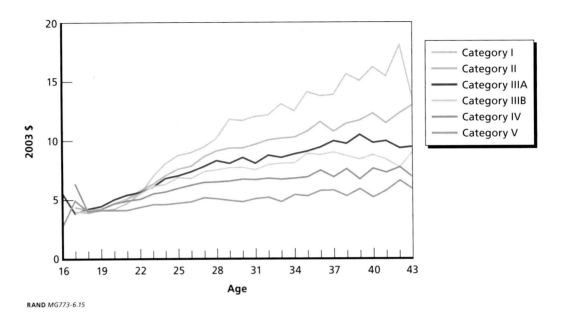

RAND *MG773-6.15*

Figure 6.16
Median Hourly Wage, by Age and Race/Ethnicity: NLSY97

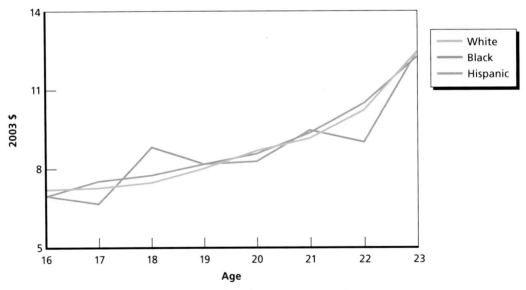

NOTE: Sample restricted to male high school graduates, Category I–II.
RAND *MG773-6.16*

Figure 6.17
Median Hourly Wage, by Age and Race/Ethnicity: NLSY79

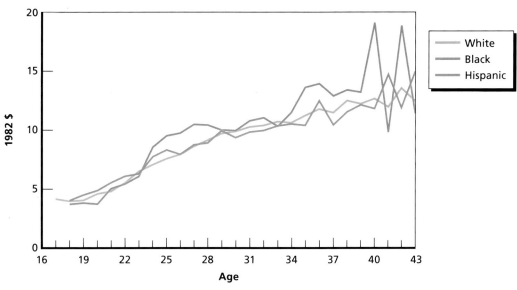

NOTE: Sample restricted to male high school graduates, Category I–II.
RAND *MG773-6.17*

Table 6.9
The Effect of Race/Ethnicity and AFQT on Log Hourly Wages and Annual Earnings

	Ln (Wage)		Ln (Earnings)	
	(1)	(2)	(3)	(4)
Black	−0.23	−0.08	−0.89	−0.56
	(0.02)**	(0.02)**	(0.08)**	(0.09)**
Hispanic	−0.06	0.01	−0.18	−0.02
	(0.02)**	(0.02)	(0.08)*	(0.09)
AFQT		0.01		0.03
		(0.00)**		(0.00)**
AFQT^2		−0.00		−0.00
		(0.00)**		(0.00)**
Constant	2.12	1.67	9.66	8.55
	(0.09)**	(0.09)**	(0.39)**	(0.43)**
Observations	36,697	36,697	39,799	39,799
R^2	0.13	0.15	0.05	0.06

SOURCE: NLSY79.

NOTES: Dependent variable in (1) and (2) is the logarithm of hourly wages. Dependent variable in (3) and (4) is logarithm of annual earnings. Sample restricted to male high school graduates age 25 and older. Excluded race/ethnicity category is white. All regressions control for age, education, and year. Robust standard errors are in parentheses.

* Significant at 5 percent.

** Significant at 1 percent.

are about 23 percent less than white males. Hispanic males earn hourly wages that are about 6 percent less than white males. Column (2) reports results that further condition on AFQT. The inclusion of AFQT in the model reduces the black-white wage gap to 8 percentage points and virtually eliminates the wage gap between whites and Hispanics.[18] Further examination of these results reveals that it is only in lower-AFQT categories that a wage gap between blacks and whites persists. In Table 6.10, we report the coefficient on black and Hispanic from models parallel to those reported in Table 6.9, but where the sample is now divided according to AFQT category (V, IV, IIIB, IIIA, II, and I). As can be seen, the black-white wage gap is greatest among Category IV and IIIB individuals. Indeed, only for the Category IV and IIIB sample can we reject the null hypothesis that being black has no independent effect on wages. The coefficient on Hispanic is statistically insignificant in every sample, but for Category V, where Hispanics actually earn higher hourly wages than do whites.

The same regression analysis employing the logarithm of annual earnings as the dependent variable shows larger differences between whites and Hispanics, on the one hand, and blacks, on the other. Conditional on educational attainment and AFQT, Hispanics and whites have statistically identical annual earnings, whereas the mean earnings of black males lag 56

Table 6.10
The Effect of Race/Ethnicity and AFQT on Log Hourly Wages and Annual Earnings, by AFQT

	AFQT					
	V	IV	IIIB	IIIA	II	I
Ln(Wage)						
Black	0.05	−0.16	−0.10	−0.05	0.00	−0.03
	(0.05)	(0.03)**	(0.04)*	(0.05)	(0.05)	(0.19)
Hispanic	0.17	−0.06	0.04	−0.03	−0.00	0.09
	(0.06)**	(0.04)	(0.04)	(0.06)	(0.05)	(0.15)
Observations	3,964	8,895	6,497	5,026	9,534	2,781
R^2	0.04	0.03	0.06	0.06	0.10	0.13
Ln(Earnings)						
Black	−0.36	−0.89	−0.63	−0.36	−0.23	0.47
	(0.31)	(0.16)**	(0.20)**	(0.25)	(0.24)	(0.36)
Hispanic	−0.03	−0.21	−0.04	0.03	0.16	−0.43
	(0.41)	(0.16)	(0.18)	(0.21)	(0.13)	(0.57)
Observations	4,347	9,840	7,089	5,372	10,099	3,052
R^2	0.03	0.03	0.02	0.04	0.03	0.05

SOURCE: NLSY79.

NOTES: Dependent variable in Panel A is the logarithm of hourly wage. Dependent variable in Panel B is logarithm of annual earnings. Sample restricted to male high school graduates age 25 and older. Excluded race/ethnicity category is white. All regressions control for age, education, AFQT, $AFQT^2$, and year. Robust standard errors are in parentheses.

* Significant at 5 percent.

** Significant at 1 percent.

---

[18] The results with respect to blacks are similar to those reported by Neil and Johnson (1996), who also use the NLSY79. Shapiro (1984) reports similar results with respect to Hispanics using data from the first few waves of the NLSY79, although not controlling for AFQT.

percent behind those of white males (Table 6.9). Again, we see that this difference in earnings is most prominent (and statistically significant) among males with lower AFQT scores (Table 6.10). Annual earnings are a function of both wages and hours worked. Given the wage results just presented, it is likely that these large earnings differences between whites and blacks are attributable, at least in part, to fewer hours of work among blacks.

We conclude this section with an observation from Table 6.4. In that table, we noted that the propensity to apply for military service generally declines with AFQT for whites, blacks, and Hispanics alike. But the specific propensities by AFQT vary across the three groups. This variation across race/ethnicity is most notable among Category II youth. Table 6.5 shows the propensity to apply relative to Category IIIB youth by AFQT and race/ethnicity. For whites and blacks, the propensity to apply among Category II youth is about 35 percent less than that of Category IIIB youth. For Hispanics, though, the propensity to apply among Category II youth is about 60 percent less than that of Category IIIB youth. Thus, it would appear that Category II Hispanics are less inclined to apply, relative to Category I youth, than are Category II white and black youth.[19] Category IIIA Hispanics also appear to be relatively less likely to apply for military service. The relatively low propensity to apply for military service among Hispanic Category II and IIIA youth is puzzling since their labor market opportunities appear to be quite similar to those of Category II whites and blacks. Evidently, other factors, perhaps differences in how these ethnic and racial groups (both applicants and their families) perceive the military, are causing relatively more-qualified Hispanic applicants to seek non-military employment.

Conclusions

Increasing the supply of Hispanics to the military poses a number of challenges to DoD. First, a large percentage of Hispanic youth either have not graduated from high school or score in Category IV or V of the AFQT distribution. Our analyses suggest that family background characteristics are important in determining educational attainment, but DoD recruiters and recruiting advertising could nonetheless inspire some Hispanic youth to finish high school who would otherwise not do so. Second, the largest pool of eligible Hispanic youth is, in effect, Category IIIB.[20] DoD already recruits a disproportionate number of Hispanics from this group, and pulling even more Hispanics from the bottom of the AFQT distribution could be problematic. Third, the supply of so-called "high quality" (Category I-IIIA high school graduates) Hispanic youth is small. Moreover, these Hispanic youth have strong civilian labor market opportunities. The armed forces face particularly stiff competition from the civilian sector for these Hispanic youth.

This descriptive analysis has shown that differences in high school graduation rates and AFQT scores can explain much of the observed difference in subsequent educational attain-

---

[19] The AFQT distribution of applicants and enlistees changed following September 11, 2001, with Category I and II applicants and enlistees becoming relatively more common. However, the differences in relative propensities to apply across racial/ethnic groups remain. In particular, compared with whites and blacks, Hispanic Category II and IIIA youth were less likely to apply relative to Category IIIB youth.

[20] Strictly speaking, Category II Hispanics are the largest group of eligible recruits, but, as seen in Table 6.2, DoD accepts very few Category IV Hispanic youth (only 2 percent of all Hispanic recruits in 2000). Excluding Category IV and V youth, the largest pool of Hispanic high school graduates is Category IIIB (see Table 6.2).

ment and labor market outcomes of whites, blackstsrdAesa, and Hispanics. But some differences remain. Hispanics, for example, appear to be more likely to attend a two-year college than either whites or blacks. Category I and II blacks are less likely to be employed than comparable whites and Hispanics; they also work fewer hours and have shorter tenure. In addition, Category IV and IIIB blacks appear to earn lower wages than comparable whites and Hispanics.

CHAPTER SEVEN

Policy Implications

Hispanics are a growing segment of the youth population, yet are underrepresented among military recruits. Hispanic underrepresentation is puzzling, considering that Hispanic young people are more likely than other groups to express a positive attitude toward the military.

A widely cited reason is Hispanics' below-average rate of graduation from high school, combined with the services' preference for recruits with high school diplomas. But other, less studied, factors may also contribute. This project was designed to analyze the factors that lead to the underrepresentation of Hispanic youth among military enlistments. Then, to help policymakers evaluate the feasibility of improving Hispanic enlistments by recruiting more intensively from among the population of youth who are qualified for service and the implications of recruiting Hispanics who are less qualified, we also analyzed both the nonmilitary opportunities available to qualified Hispanic youth and the consequences of recruiting less-qualified Hispanic youth. Our study was designed to answer three key questions:

- Which entry standards disqualify Hispanics more than other groups?
- If recruiting standards were relaxed, what would be the effect on military performance—specifically, career outcomes?
- What actions could be taken to increase Hispanic enlistments given Hispanics' civilian and educational opportunities?

Although this study focuses primarily on Hispanics, we also include information for other groups, including blacks and whites, when the data allow.

In terms of entry standards, we found that the major characteristics that disproportionately disqualify Hispanic youth are lack of a high school diploma, lower AFQT scores, and being overweight. Other factors (such as exceeding the number of allowable dependents) also contribute, but these three are the most significant. The three factors tend to correlate positively. In other words, candidates who lack a high school diploma are more likely to score poorly on the AFQT; similarly, overweight candidates are also more likely to score poorly on the AFQT.

However, another key finding is that, in terms of military outcomes (such as retention rates and promotion speed), the effect of ethnicity is more significant than the effect of standards. Hispanics usually have higher retention rates and faster promotion speeds than their white counterparts. (The only exception is promotion in the Navy.) Moreover, lower-quality Hispanic recruits (i.e., those with somewhat lower AFQT scores, less education, and greater weight) often compare well to their higher-quality white counterparts. These results suggest that these Hispanic recruits may be better matched to the military's demands and lifestyle, or have better opportunities in the military than in the civilian sector.

The results for black recruits are roughly the same as for Hispanic recruits. Retention rates are higher for black recruits in the Army and Air Force, and promotion speeds are faster in the Army than for similar white recruits. Similarly, we find that lower-quality black recruits have higher retention rates than white recruits who are higher quality or who meet or exceed entry standards. Thus, like Hispanic recruits, black recruits seem to be better suited to the military or to have better opportunities in the military than in the civilian sector.

One implication of these results is that lowering AFQT, education, and weight standards while targeting the enlistment of Hispanics and blacks would have little effect on retention and promotion rates. In fact, the results can be stated more strongly. Faced with the choice of otherwise identical whites, Hispanics, and blacks, the military should choose Hispanics and blacks. Moreover, faced with the choice of somewhat higher-quality whites versus somewhat lower-quality Hispanics and blacks, the military should choose the Hispanics and blacks. Hispanic and black recruits will yield more man-years (due to lower attrition and higher retention) and more leaders. On the other hand, lowering standards without targeting would hurt retention and promotion rates.

We make the above statements based solely on the results of the numerical analysis. We are, of course, aware that a policy of overt and explicit discrimination would not and should not be tolerated in reality. We do not advocate such a policy in any way. However, recruiting commands and recruiters make decisions every day about how to allocate recruiting resources. The results of our study suggest that it would be worthwhile to direct more recruiting resources to black and Hispanic markets. Such recruiting resources might include advertising dollars, assignment of recruiters, and guidance to recruiters about successful recruiting strategies.

Weight is a major area in which Hispanics are less likely to meet enlistment eligibility standards. Again, the effect of ethnicity is more significant than the effect of standards. Overweight minorities have higher retention rates than white recruits who are near the weight standard. In the case of the Marine Corps, the differences are quite large. For example, the four-year retention rate of a Hispanic Marine Corps recruit who is 20 pounds overweight is 12 percentage points above the four-year retention rate for white recruits who are within five pounds of the standard.

While the weight standard disproportionately screens out Hispanics, it is a nontrivial problem for all ethnic groups. Approximately 25–35 percent of young adult men and 50–60 percent of young adult women, regardless of race or ethnicity, would fail the weight standards of at least one branch of service. (The comparable numbers for Hispanics are about 31 percent for men and about 55 percent for women.) In addition, the alarming trend in childhood obesity has been widely acknowledged. Young people weigh more at younger ages. Thus, policies that seek to address the failure to meet weight standards will affect all race and ethnicity groups, not just Hispanics. Furthermore, such policies will grow in importance, because recent trends suggest weight will become a larger health problem in the future.

Relaxing the weight standards would not result in a one-for-one increase in the supply of recruits. As noted earlier, a number of characteristics are correlated. Being overweight is correlated with other disqualifying characteristics, such as lower AFQT score. However, relaxing the weight standard would increase the pool of eligible candidates, even if not on a one-for-one basis. Five policy options might be explored.

First, the military could create programs to help potential recruits reduce their weight before entry. Currently, if candidates cannot meet the weight standards on the day they are processed for enlistment, they are required to restart the process at a later date. Instead of reprocessing, it may be useful to enroll such candidates in a weight reduction program while in the delayed entry program. However, the services should keep in mind that this could be costly, and further study would be needed to evaluate whether the weight reductions could be sustained once the recruits are enlisted and serving on active duty. If recruits cannot sustain an acceptable weight, the result will be higher attrition. This concern should be balanced with the problem of turning away candidates who are otherwise qualified and interested, especially in a difficult recruiting environment.

A second alternative might be to relax the weight standards at entry. The Army and Marines currently have different weight standards for new accessions and active duty. These standards differ by two to three pounds. These new accession standards could be revised and put in place for all branches. Slightly overweight but otherwise qualified individuals would be allowed to enter service and be expected to take off excess weight in boot camp/basic training. Such a policy would have to be implemented in concert with programs to help meet and sustain the weight standards.

A third alternative would be to stratify the weight requirements by job. Variation in standards currently exists across branches. The Air Force and Navy weight standards are not as stringent as the Army or Marine Corps standards, yet personnel in these services can be just as effective at their service-specific mission. Such a policy would not necessarily imply a change in fitness standards (e.g., the requirements for running and doing sit-ups and push-ups could remain the same), and incentives to exceed the minimum standard could be provided. Such incentives might include promotion points or other career-related rewards. This policy would simply admit that not all members are airborne infantrymen; they may be equally competent with different weight requirements. However, this recommendation assumes that individuals would know their occupational specialty at entry, so that the different weight standards could be applied upon in-processing. This may not be feasible for all branches of the military because individuals may not be assigned to a specialty until after basic training.

A fourth alternative might be to relax the weight standards while maintaining the strength standards. This policy is already being implemented by the Marine Corps, which has the most lenient weight standard but requires a strength test for individuals who receive a waiver. It is possible that the Marine Corps' lenient policy toward weight, coupled with its strength requirement, has allowed that service to attract recruits who are well suited to it and therefore are less likely to leave. Interestingly, the four-year Marine Corps retention rate of Hispanic recruits who are over 20 pounds overweight is not much different from the retention rates of those who are within five pounds of the standard. Apparently, the addition of the strength test weeds out those who are unlikely to complete service. The Army, too, has introduced an experimental program that allows applicants who exceed the weight standard to qualify for enlistment, conditional on having a qualified body mass index. The applicants must prove they are physically fit, based on the Harvard Step Test.

The final alternative is to recruit from healthier populations. Youth with more education (especially with some college) and noncitizens are more likely to pass weight standards. Enlistment programs that fast-track the citizenship of immigrants serving in the military and recruitment programs targeting college-bound youth may have the added benefits of improving the likelihood of youth meeting weight standards or of improving Hispanic representation among enlistments. However, the possibility of fast-tracking immigrant citizenship should be approached with caution; specifically targeting immigrants to serve in the military could produce strong political and social backlash.

Increasing the supply of Hispanics to the military poses a number of challenges to the services. For the least qualified of this population, a group that is relatively large, the research suggests that identifying and targeting the most motivated of this group offers advantages and is consistent with current efforts like the Army's Tier Two Attrition Screen program.

The relatively low AFQT scores and high weight of Hispanics reflect in large part family and neighborhood characteristics that are difficult to change in the short run. Nonetheless, the services do have some ability to affect eligibility in both areas. Alternative policies regarding weight are discussed above. To the extent that low AFQT scores reflect English language proficiency, the services could expand existing programs that target improvement in language skills.

The supply of "high quality" Hispanic youth is small. ("High quality" youth are high school graduates in the three highest AFQT groups—Categories I, II, and IIIA.) Moreover, these Hispanic youth have strong civilian labor market opportunities: The services face particularly stiff competition from the civilian sector for these young people. The services must find ways to compete with excellent civilian opportunities available to this group. The military compensation system should reward excellent performance and be competitive with compensation in the civilian sector. Moreover, the services should emphasize the nonpecuniary benefits of service, such as leadership opportunities, greater responsibility, and opportunities to serve one's country.

Finally, college—especially two-year college—seems quite important, for Hispanic youth. (Hispanics are more likely to attend a two-year college than either whites or blacks.) Since many Hispanics do not complete college, and many work while in college, more exploration is needed as to whether these individuals lack resources or have lower educational expectations. In either case, military service as part of one's educational path, and the suite of educational benefits available to those who serve, could be marketed more heavily to this group. Educational benefits are likely to be well received among high-quality Hispanics, just as they are among high-quality blacks and whites.

Educational benefits are only one of many recruiting resources. Little is known about how Hispanic and black recruits respond to educational benefits and other recruiting resources, such as enlistment bonuses, and to external factors including the Iraq war, the civilian economy, and college opportunities. Such information would be useful for developing policy options to increase the supply of Hispanic and black recruits. Given recent declines in Army enlistments among blacks and concerns about social representation, a subject for future research could be investigating which resources are the most effective with each population group.

It is common to argue today that the military can no longer recruit just the individual; the military must recruit the family as well. This may be especially true of Hispanics, whose family ties are perhaps particularly strong. Future research on the factors determining Hispanic enlistments should pay close attention to the role of the family in determining attitudes toward the military as a career.

Detailed Information on Enlistment Standards

This appendix supplements the information provided in Chapter Two by providing more detailed information on enlistment standards. Appendix B provides additional information on service policy regarding waivers.

Citizenship

As discussed in Chapter Two regarding the standard for citizenship, some individuals are permanent residents while others are conditional permanent residents. The terms used for the purposes of enlistment are defined in this appendix.[1]

- **Alien Lawfully Admitted for Permanent Residence.** Known as LPRs, green card holders, or permanent residents, they can reside in the United States indefinitely unless their immigration status is changed, and they are not require to give allegiance to the United States.
  - *Alien Registration Receipt Card.* The Immigration and Nationality Act requires adult lawful permanent resident aliens to carry an alien registration receipt card, which is INS (Immigration and Naturalization Services) Form I-551. Although those who hold this card are permanent residents, they are required to renew the card after ten years; thus, most versions of Form I-551 contain an expiration date on the front.
- **Alien Lawfully Admitted for Conditional Permanent Residence.** An alien lawfully residing permanently in the United States on a conditional basis. The only difference between permanent residents and conditional permanent residents is that the status of a conditional resident will expire two years from the date of issue unless the alien successfully petitions to have this condition removed, in which case the individual becomes a permanent resident. These aliens receive an Alien Registration Receipt Card (Form I-551) with an expiration date two years from date of issue. This status is granted to an alien and the alien's children based upon the alien's marriage to a U.S. citizen or lawful permanent resident.
- **U.S. Noncitizen National.** A person who, though not a U.S. citizen, owes permanent allegiance to the United States. Persons born in American Samoa and the Swains Island are noncitizen nationals of the United States.

---

[1] The definitions in this section are paraphrased from U.S. Navy Recruiting Command, COMNAVCRUITCOMINST 1130.8F, 2003. Although stated by the Navy, these definitions apply to all services.

Dependency Status

DoD Directive 1304.26 disqualifies applicants who are married and have more than two dependents less than 18 years of age and nonmarried applicants with any dependents less than 18 years of age.[2] The Army and Air Force standards are almost identical to that mandated by the DoD. They allow married recruits to have three dependents (a spouse and two additional dependents) before a waiver is required. The Marine Corps and Navy set more restrictive standards, requiring a waiver for married applicants with more than one dependent. Because the spouse of a married applicant is defined as a dependent, applicants enlisting in the Marine Corps or Navy who have any dependents other than their spouse require a waiver.

All services prohibit the acceptance of single parents with any dependents under the age of 18. This requirement has no exceptions and waivers are not available for single parents.

Some differences between the services' dependency requirements are found in the definitions used to classify dependents. One such difference is how each service defines when a spouse is or is not considered a dependent. The Navy classifies a spouse as a dependent "regardless of financial support or spouse's military status, unless terminated by final divorce decree" (COMNAVCRUITCOMINST 1130.8F, p. 2-32). The Army is more lenient and does not consider a spouse as a dependent if

- common law marriage has not been recognized by a civil court
- spouse is incarcerated
- spouse is deceased
- spouse has deserted the applicant
- spouse is legally or by mutual consent separated from the applicant
- the applicant or the spouse has filed for divorce (AR 601-210, p. 8).

The Marine Corps states that "a spouse, to include a common law spouse if the state recognizes such," is considered a dependent (MCO P1100.72C, p. 3-33). Air Force documentation simply states that a spouse is considered a dependent (AETC Instruction 36-2002, p. 76).

The services also differ on whom they define as a dependent child (dependents less than 18 years of age and unmarried). There are differences in when the services define natural children, stepchildren, and adopted children as dependents, as shown in Table A.1.

In addition to children, each service considers other people whom applicants support financially or otherwise as dependents, but there are slight differences in how the different services define these dependents.

- The Army includes "any person for whom the applicant is responsible for his or her financial or custodial care" as a dependent. This includes "any person living with the applicant who is, by law or in fact, dependent upon the applicant for support; or not living with the applicant and dependent upon the applicant for over one-half of his or her support" (AR 601-210, p. 8).
- The Air Force includes "any person over the age of 18 incapable of self-care for whom the applicant or spouse has assumed responsibility for care" (AETC Instruction 36-2002, p. 76).

---

[2] The directive also gives the services the right to give waivers for particularly promising entrants.

Table A.1
Definitions of Children as Dependents, by Service

Service	A natural child is considered a dependent:	An adopted child is considered a dependent:	A stepchild is considered a dependent:
Army	if the child has not been placed in custody of other parent or adult by court order	if the child is living with the applicant	if the child is living with the applicant
Air Force	if the child has not been legally adopted by another adult	if the child has not been legally adopted by another adult	whether or not the child is living with applicant
Navy	if the child has not been legally adopted by another adult	if the child has not been legally adopted by another adult	if the applicant's spouse has custody of the child
Marine Corps	if the child has not been legally adopted by another adult	regardless of whether or not the applicant has custody of the child	if the child is living with the applicant

- The Navy states: "aside from children, spouse or stepchildren, dependency depends on whether the applicant is providing financial support to the 'dependent'" (COM-NAVCRUITCOMINST 1130.8F, p. 2-32).
- The Marine Corps includes "any parent or other person(s) who is/are, in fact, dependent on the applicant for more than one-half of their support" as a dependent (MCO P1100.72C, p. 3-33).

Aptitude

The following information is quoted directly from the following web site: http://usmilitary. about.com/library/milinfo/blafqtscore.htm (as of March 19, 2004).

Only four areas of the ASVAB test are used to compute the overall ASVAB score, also known as the AFQT (Armed Forces Qualification Test) score.

The four areas of the ASVAB used to compute the AFQT score are: Word Knowledge (WK), Paragraph Comprehension (PC), Arithmetic Reasoning (AR), and Mathematics Knowledge (MK).

To determine the AFQT "raw score," first you have to compute your Verbal Expression (VE) score:

VE (Verbal Expression) = Scaled Score of WK + PC. To get the VE scaled score, add your Word Knowledge (WK) & Paragraph Comprehension (PC) score together, then use the below table:

Table A.2
Computation of the Raw AFQT Score

WK + PC	VE Score	WK + PC	VE Score
0–3	20	28–29	42
4–5	21	30–31	44
6–7	22	32–33	45
8–9	22	34–35	47
10–11	25	36–37	49
12–13	27	38–39	50
14–15	29	40–41	52
16–17	31	42–43	54
18–19	32	44–45	56
20–21	34	46–47	58
22–23	36	48–49	60
24–25	38	50	62
26–27	40		

The overall ASVAB Score (AFQT Score) is a "percentile score." The AFQT Raw Score is computed with the formula AFQT = 2VE + AR + MK.

You can't use the AR and MK score shown on your ASVAB Score Sheet. The Score Sheet shows "number correct" for your AR and MK Scores, because "number correct" is what is used for job qualifications. However, the military does not use this same score when computing the AFQT. They use the "weighted scores" of the ASVAB subtests for AR and MK. Harder questions in these areas get more points than easier questions. The "weighted scores" for AR and WK are not listed on the ASVAB score sheet given to you after the test, and [this web site's author knows] of no way to retrieve that information from DOD.

The "raw score" is then compared with the above chart to determine overall ASVAB percentile score.

Moral Character

Each service has individual moral character standards.

Army. The Army separates crimes into four categories: (1) typical minor traffic violations, (2) typical minor non-traffic violations, (3) typical misdemeanors, (4) typical felonies offenses. For examples of crimes that fit into these different categories see Table A.3. The moral standards set by the Army require a waiver for any applicant who has

- received a civil court conviction or other adverse dispositions for six or more minor traffic offenses where the fine was $100 or more per offense.
- received three or more civil convictions or other adverse dispositions for minor non-traffic offenses
- received two, three, or four civil convictions or other adverse dispositions for a misdemeanor offense. (See below for DUI/DWI [driving under the influence/driving while intoxicated])

- received a total of three civil convictions or other adverse dispositions for a combination of minor non-traffic and misdemeanor (one misdemeanor and two minor non-traffic). (See below for DUI/DWI)
- received one conviction or adverse disposition for a DUI/ DWI (See below for more than one DUI/DWI offense)
- received a conviction or other adverse disposition for a felony offense
- received two convictions or adverse dispositions for driving while intoxicated, under the influence, or while impaired due to substance abuse, alcohol, drugs, or any other condition that affected judgment or driving ability. Consider without regard to technical/legal definition or term used by the state, county, or country in which the applicant committed the offense (AR 601-210, p. 24).

In addition to the waivers required above, applicants with the following are disqualified and not eligible for Army waivers:

- convictions or other adverse dispositions for five or more misdemeanors preceding application for enlistment
- three or more convictions or other adverse dispositions for driving while intoxicated, drugged, or impaired in the five years preceding application for enlistment
- subject of initial civil court conviction or adverse disposition for more than one felony offense
- civil conviction of a felony with any of the following:
  - three or more offenses (convictions or other adverse dispositions) other than traffic
  - juvenile felony offenses who have had no offenses within five years of application for enlistment may be considered for a waiver in meritorious cases (AR 601-210, p. 29).

Air Force. The Air Force breaks crimes into five categories. For examples of crimes that fit into the different categories, see Table A.4. The categories are as follows:

- Category 1 offenses are major offenses and can be waived for entrance into the Air Force only by the Air Force Recruiting Service (AFRS) commander or vice commander.
- Category 2 offenses are also major offenses, but of a less serious nature. These offenses can be waived for entrance into the Air Force only by recruiting group commanders or deputy commanders.
- Category 3 offenses are serious offenses. A conviction for these offenses can be waived only by a recruiting squadron commander.
- Category 4 offenses are less serious offenses. Two convictions in the last three years or three or more convictions in a lifetime require a waiver by a recruiting squadron commander for entry into the Air Force.
- Category 5 offenses are traffic offenses. Six or more convictions in any 365-day period in the last three years require a recruiting squadron commander's waiver approval for entry into the Air Force. Again, quality and the good of the Air Force must be the overriding factor in the submission and approval of moral waivers (AETC Instruction 36-2002, p. 77).

Navy. The Navy divides crimes into four categories: (1) minor traffic violations, (2) minor non-traffic/minor misdemeanors, (3) non-minor misdemeanors, (4) felonies. For examples of crimes that fit into these different categories, see Table A.5.

- Applicants with six or more category 1 violations, three to five category 2 violations, and any category 3 or 4 violations require a waiver.
- Applicants with six or more category 2 and 3 violations or multiple felonies are disqualified from enlisting and are not eligible for a waiver.

Marine Corps. The Marine Corps divides crimes into six categories: (1) minor traffic offenses, (2) serious traffic offenses, (3) Class 1 minor non-traffic offenses, (4) Class 2 minor non-traffic offenses, (5) serious offenses, and (6) felony offenses. For examples of crimes that fit into these different categories, see Table A.6. Applicants who have a conviction, adverse adjudication, or have served or been credited a term of incarceration for

- Five or more minor traffic; two or more serious traffic; four or more Class 1 minor non-traffic; and two to five Class 2 minor non-traffic offenses require a waver from the recruiting station.
- Six to nine Class 2 minor non-traffic offenses and/or three to five serious offenses require a waiver from the Marine Corps District.
- One felony offense requires a waiver from the Recruiting Region (MCO P1100.72C, p. 3-131).

Additionally, applicants are ineligible for the Marine Corps if they have ten or more Class 2 minor non-traffic offenses; or six or more serious offenses; or more than one felony offense.

Table A.3
Categories of Crimes: Army

Typical minor traffic offenses: a. Blocking or retarding the traffic. b. Bicycle ordinance violation. c. Crossing yellow line, driving left of center. d. Disobeying traffic lights, signs, or signals. e. Driving on shoulder. f. Driving uninsured vehicle. g. Driving with blocked vision/tinted window. h. Driving with expired plates or without plates. i. Driving without license or with suspended or revoked license. j. Driving without registration or with improper registration. k. Driving wrong way on one-way street. l. Failure to appear for traffic violations. m. Failure to comply with officer's directive. n. Failure to have vehicle under control. o. Failure to signal. p. Failure to stop or yield to pedestrian. q. Failure to submit report after accident. r. Failure to yield right-of-way. s. Faulty equipment, such as defective exhaust, horn, lights, mirror, muffler, signal device, steering device, tail pipe, or windshield wipers. t. Following too closely. u. Hitchhiking. v. Improper backing, such as backing into intersection or highway, backing on expressway, or backing over crosswalk. w. Improper blowing of horn. x. Improper passing, such as passing on right, passing in no-passing zone, passing stopped school bus, or passing a pedestrian in crosswalk. y. Improper turn. z. Invalid or unofficial inspection sticker, failure to display inspection sticker. aa. Jaywalking. ab. Leaving key in ignition. ac. License plates improperly displayed or not displayed. ad. Operating overloaded vehicle. ae. Racing, dragging, or contest for speed. af. Reckless, careless or imprudent driving (considered a traffic offense when the fine is less than $250.00 and there is no confinement). Court costs are not part of a fine. ag. Seat belt/child restraint violation. ah. Skateboard/roller skate violations. ai. Speeding. aj. Spilling load on highway. ak. Spinning wheels, improper start, zigzagging, or weaving in traffic. al. Violation of noise control ordinance.

Typical minor non-traffic offenses: a. Assault, fighting or battery (no confinement) (less than $250.00 fine). b. Carrying concealed weapon (other than firearm); possession of brass knuckles. c. Criminal or malicious mischief (less than $250.00 fine). d. Curfew violation. e. Damaging road signs. f. Discharging firearm through carelessness. g. Discharging firearm within municipal limits. h. Disobeying summons, failure to appear. i. Disorderly conduct; creating disturbance, boisterous conduct. j. Disturbing the peace. k. Drinking alcoholic beverages on public transportation. l. Drunk in public; drunk or disorderly. m. Dumping refuse near highway. n. Illegal betting or gambling; operating illegal handbook, raffle, lottery, or punch board; matching cockfight. o. Jumping turnstile (to include those states that adjudicate jumping a turnstile as petty larceny). p. Juvenile adjudications such as beyond parental control, incorrigible, runaway, truant, or wayward. q. Killing domestic animal. r. Liquor; unlawful manufacture, sale, possession, or consumption in public place. s. Littering. t. Loitering. u. Nuisance, committing. v. Poaching. w. Purchase, possession, or consumption of alcohol beverages or tobacco products by minor. x. Removing property from public grounds. y. Removing property under lien. z. Robbing orchard. aa. Shooting from highway. ab. Shooting on public highway. ac. Solicitation (other than for commission of a felony, prostitution or sexual offense). ad. Throwing glass or other material in roadway. ae. Trespass (non-criminal). af. Unlawful assembly. ag. Using or wearing unlawful emblem/identification. ah. Vagrancy. ai. Vandalism (less than $250.00 fine). aj. Violation of fireworks law. ak. Violation of fish and game laws. al. Violation of leash laws

Typical misdemeanors: a. Altered driver's license or identification. b. Assault, fighting or battery (confinement imposed), or a fine of $250.00 or more. c. Auto burglary/tampering with an auto (value less than $250.00). Are typical offenses where parts or items were removed from a vehicle. (if value exceeds $250.00, offense becomes a felony.) d. Being in place where narcotics or habit-forming drugs are being used. e. Check, worthless, making or uttering, with intent to defraud or deceive (less than $250.00). f. Conspiring to commit misdemeanor. g. Criminal contempt of court. h. Contributing to delinquency of minor. i. Crimes against the family (nonpayment of child support) if tried in a criminal court. j. Criminal or malicious mischief (fine of $250.00 or more). k. Desecration of American flag. l. Desecration of grave. m. Driving while drugged or intoxicated, or driving while ability impaired (single offense). n. Failure to register with Selective Service. o. Failure to stop and render aid after accident. p. False bomb threat. q. Glue sniffing. r. Harassment. s. Illegal or fraudulent use of a credit card, bank card, or automated teller (ATM) card (value less than $250.00). t. Indecent exposure. u. Indecent, insulting, or obscene language communicated directly or by telephone to another person. v. Joy riding. w. Larceny or conversion (value of less than $250.00). x. Leaving scene of an accident or hit and run. y. Looting. z. Permitting a DUI. aa. Paint/chemical sniffing. ab. Reckless driving, careless, or imprudent. (Considered a misdemeanor when the fine is $250.00 or more or when confinement is involved; otherwise, considered a minor traffic offense). ac. Reckless endangerment. ad. Resisting arrest (eluding police, fleeing). ae. Selling or leasing weapons. af. Stolen property, knowingly received (value less than $250.00). ag. Trespass-criminal. ah. Unlawful carrying of firearms or carrying concealed firearm. ai. Unlawful entry. aj. Unlawful use of long-distance telephone lines. ak. Use of telephone to abuse, annoy, harass, threaten, or torment another. al. Vandalism (fine of $250.00 or more). am. Violation of probation. an. Willfully discharging firearm so as to endanger life; shooting in public place.

Typical felony offenses: a. Aggravated assault, assault with dangerous weapon, assault intentionally inflicting great bodily harm, or assault with intent to commit a felony. This also includes child, parental, or spouse abuse. b. Arson. c. Attempt to commit a felony. d. Breaking and entering. e. Bribery. f. Burglary, (burglary tools, possession of). g. Carnal knowledge of a minor h. Check, worthless, making or uttering, with intent to defraud or deceive ($250.00 or more). i. Conspiring to commit a felony. j. Criminal libel. k. Driving while drugged or intoxicated, or driving while ability impaired (2 or more offenses). l. Extortion. m. Forgery; knowingly uttering or passing forged instrument. n. Graft. o. Illegal/fraudulent use of a credit card, bank card, or automated (ATM) card (value of $250.00 or more). p. Indecent acts or liberties with a minor. q. Indecent assault. r. Kidnapping or abducting, to include parental kidnapping of a child(ren). s . Larceny; embezzlement; conversion (value of $250.00 or more). t. Mail matter; abstracting, destroying, obstructing, opening, secreting, stealing, or taking. u. Mails; depositing obscene or indecent matter. v. Manslaughter. w. Mis-prison of felony. x. Murder. y. Narcotics or habit-forming drugs; wrongful possession or use. z. Negligent/vehicular homicide. aa. Pandering. ab. Perjury or subornation of perjury. ac. Public record; altering, concealing, destroying, mutilating, obligation, or moving. ad. Rape, sexual abuse, sexual assault, criminal sexual abuse, incest. ae. Riot. af. Robbery. ag. Sodomy. ah. Stolen property, knowingly received (value $250.00 or more). ai. Solicitation or Prostitution.

Table A.4
Categories of Crimes: Air Force

Category 1 Moral Offenses: This list of offenses is a guide. Consider violations of a similar nature of seriousness as a category 1 offense. A conviction or adverse adjudication of one or more of these offenses is disqualifying for entry into the Air Force. The procurement source commander approves waivers to these offenses: Aggravated assault: With a dangerous weapon, intentionally inflicting great bodily harm, with intent to commit a felony (adjudicated as adult only). Bribery (adjudicated as adult only).Burglary (adjudicated as adult only). Carnal knowledge of a child under 16. Draft evasion. Drugs: Use, possession, trafficking, sale, or manufacture of an illegal or illicit drug (except for marijuana use or possession - see category 2). Extortion (adjudicated as adult only). Indecent acts or liberties with a child under 16, molestation. Kidnapping, abduction. Manslaughter. Murder. Perjury (adjudicated as adult only). Rape. Robbery (adjudicated as adult only)

Category 2 Moral Offenses: This list of offenses is a guide. Consider violations of a similar nature or seriousness as a category 2 offense. In doubtful cases, treat the offense as a category 2 offense when the maximum possible confinement under local law exceeds one year. Conviction or adverse adjudication of one or more of these offenses is disqualifying for entry into the Air Force. Waivers to these offenses may be approved by the next lower level of command below the procurement source: Arson. Aggravated assault: With a dangerous weapon, intentionally inflicting great bodily harm, with intent to commit a felony (adjudicated as juvenile only). Attempting to commit a felony. Breaking and entering a building with intent to commit a felony. Burglary (adjudicated as juvenile only). Bribery (adjudicated as juvenile only). Carrying a concealed firearm or unlawfully carrying a firearm. Carrying a concealed weapon (other than firearm), possession of brass knuckles. Child pornography offenses. Conspiring to commit a felony. Criminal libel. DUI/DWUI/DWI (driving under the influence, while intoxicated, or impaired by drugs or alcohol). Embezzlement. Extortion (adjudicated as juvenile only). Forgery: Knowingly uttering or passing forged instrument (except for altered identification for purchase of alcoholic beverages). Grand larceny. Grand theft. Housebreaking. Indecent assault. Involuntary manslaughter. Leaving the scene of an accident (hit-and-run) involving personal injury. Lewd, licentious, or lascivious behavior. Looting. Mail or electronic emissions matters: Abstracting, destroying, obstructing, opening, secreting, stealing or taking. Mail: Depositing obscene or indecent matter. Maiming or disfiguring. Marijuana: Simple possession or use. Negligent homicide. Perjury (adjudicated as juvenile only). Public record: Altering, concealing, destroying, mutilating, obliterating, or removing. Riot. Robbery (adjudicated as juvenile only). Sedition or soliciting to commit sedition. Selling, leasing, or transferring weapon to a minor or unauthorized individual. Sexual harassment. Willfully discharging firearms so as to endanger life or shooting in public place.

Category 3 Moral Offenses: This list of offenses is a guide. Consider violations of a similar nature as category 3 offenses (including boating, aviation, and similar recreational vehicular offenses). In doubtful cases, treat the offense as a category 3 offense when the maximum possible confinement under local law exceeds 4 months but no more than one year. Conviction or adverse adjudication of one or more of these offenses is disqualifying for entry into the Air Force. Waivers to these offenses may be approved by the lowest level of command in the procurement source: Adultery. Assault (simple). Breaking and entering a vehicle. Check: Insufficient funds (amount of check over $50, worthless, or uttering with intent to defraud or deceive). Conspiring to commit misdemeanor. Contempt of court (includes nonpayment of child support or alimony required by court order).

Category 4 Moral Offenses: This list of offenses is a guide. Consider traffic violations that are treated as serious by law enforcement agencies as category 4 offenses (including boating, aviation, and similar recreational vehicular offenses). In doubtful category 4, non-traffic cases, treat similar offenses as category 4 offenses when the maximum possible confinement under the local law is 4 months or less. Two convictions or adverse adjudications in the last 3 years, or three or more convictions or adverse adjudications in a lifetime is disqualifying for entry into the Air Force. Waivers to these offenses may be approved by the lowest level of command in the procurement source: Abusive language under circumstances to provoke breach of peace. Altered identification when intent is to purchase alcoholic beverages. Careless or reckless driving. Check ($50 or less, insufficient funds or worthless). Curfew violation. Committing or creating nuisance. Damaging road signs. Disorderly conduct, creating disturbance or boisterous conduct, disturbing the peace. Driving with suspended or revoked license or without license. Failure to comply with officer's direction. Failure to appear, comply with judgment, answer or disobey summons. Fare evasion (includes failure to pay turnstile fees). Fighting, participating in a brawl. Illegal betting or gambling: Operating illegal handbook, raffle, lottery, punch board or watching a cockfight. Juvenile noncriminal misconduct: Beyond parental control, incorrigible, runaway, truant or wayward. Liquor or alcoholic beverages: Unlawful possession or consumption in a public place. Littering or dumping refuse on or near highway or other prohibited place. Loitering. Possession of indecent publications or pictures (other than child pornography). Purchase, possession, or consumption of alcoholic beverages by a minor. Racing, drag racing, contest for speed. Shoplifting, larceny, petty larceny, or theft (committed under age 14 and value of stolen goods is $50 or less). Trespass on property. Unlawful assembly. Vagrancy. Vandalism, defacing or injuring property. Violation of fireworks law. Violation of fish and game laws.

Table A.4—Continued

Category 5 Moral Offenses: This list of offenses is a guide. Consider offenses of a similar nature (including boating, aviation, and similar recreational vehicular offenses) and traffic offenses treated as minor by local law enforcement agencies, as Category 5 offenses. However, careless or reckless driving are considered category 3 offenses. If the offense is for parking tickets, count and document only tickets written by law enforcement officers for parking in prohibited zones, regardless of location. Do not count or document any overtime parking tickets. Do not count any parking tickets issued by private security firms, campus police, etc. Conviction or adverse adjudication of six or more category 5 offenses in a 365-day period in the last three years is disqualifying for entry into the Air Force. Waivers to these offenses may be approved by the lowest level of command in the procurement process: Blocking or retarding traffic. Crossing yellow line, drifting left of center. Disobeying traffic lights, signs, or signals. Driving on shoulder. Driving uninsured vehicle. Driving with blocked or impaired vision. Driving with expired plates or without plates. Driving without license in possession. Driving without registration or with improper registration. Driving wrong way on one-way street. Failure to display inspection sticker. Failure to have vehicle under control. Failure to keep right or in proper lane. Failure to signal. Failure to stop or yield to a pedestrian. Failure to yield right-of-way. Faulty equipment (defective exhaust, horn, lights, etc., illegal window tint). Following too close. Improper backing. Improper blowing of horn. Improper passing. Improper turn. Improper parking (does not include overtime parking). Invalid or unofficial inspection sticker. Leaving key in the ignition. License plates improperly displayed or not displayed. Operating overloaded vehicle. Playing vehicle radio or stereo too loud (noise or sound pollution). Speeding (contest for speed, racing or drag racing is category 4 offense). Spinning wheels, improper start. Seat belt violation.

Table A.5
Categories of Crimes: Navy

Chart A—Minor Traffic Violations: Blocking or retarding traffic. Careless driving (when not treated as reckless driving). Crossing yellow line; driving left of center line. Disobeying traffic lights, signs, or signals. Driving on shoulder. Driving uninsured vehicle. Driving with blocked vision. Driving with expired plates or without plates. Driving without license in possession. Driving without registration or with improper registration. Driving wrong way on one-way street. Failure to comply with officer's directives. Failure to have vehicle under control. Failure to keep to right or in line. Failure to signal. Failure to submit report following accident. Failure to yield right-of-way. Faulty equipment (such as, defective exhaust, horn, lights, mirror, muffler, signal device, steering device, tailpipe, or windshield wipers). Following too closely. Improper backing; backing into intersection or highway; backing over crosswalk. Improper blowing of horn. Improper parking: such as, restricted area, fire hydrant; double parking, overtime parking. Improper passing: such as, passing on right, in no-passing zone; passing parked school bus, pedestrian in crosswalk (when not treated as reckless driving). Improper turn. Invalid or unofficial inspection sticker; failure to display inspection sticker. Leaving key in ignition. License plate improperly displayed or not displayed. Operating overloaded vehicle. Speeding (when not treated as reckless driving). Spinning wheels; improper start, zigzagging; or weaving in traffic (when not treated as reckless driving). Note 1. An all-inclusive list of minor traffic offenses valid for all States would be impracticable. The above list is intended as a guide. Offenses of similar nature and traffic offenses treated as minor by local law enforcement agencies should be treated as minor.

Chart B—Minor Non-Traffic Violations/Minor Misdemeanors: Abusive language under circumstances to provoke breach of peace. Carrying concealed weapon (other than firearm); possession of brass knuckles. Check, worthless, making or uttering, with no intent to defraud or deceive ($100 or less.) Curfew violation. Damaging road signs. Desecration of grave. Discharging firearm through carelessness. Disobeying summons. Disorderly conduct; creating disturbance; boisterous conduct. Disturbing peace. Driving without a license or with suspended or revoked license. Drinking in public. Drunk in public; drunk and disorderly. Dumping refuse near highway. Failure to appear. Fair/toll evasion. Illegal betting or gambling; operating illegal handbook, raffle, lottery, punch board; watching cockfight. Juvenile non-criminal misconduct; beyond parental control; incorrigible; runaway; truant; or wayward. Liquor: unlawful manufacture, sale, possession, or consumption in public place. Littering. Loitering. Malicious mischief. Nuisance, committing. Poaching. Possession of alcohol by minor. Possession of cigarettes by minor. Possession of indecent publications or pictures (Contact CNRC, Code 017 for determination). Probation Violation (Contact CNRC Code 017). Public urination. Purchase, possession, or consumption of alcoholic beverages by minor. Racing, dragging, contest for speed (when not treated as reckless driving). Removing property under lien. Removing property from public grounds. Robbing orchard. Trespass to property. Unlawful assembly. Use of false ID to buy alcohol. Using or wearing unlawful emblem. Vagrancy. Vandalism: injuring or defacing public property or property of another; shooting out streetlights. Violation of fireworks laws. Violation of fish and game laws. Note 1. The above list is intended as a guide. Violations of a similar nature should be treated as minor violations. In doubtful cases the following rule should be applied: If the maximum confinement under local law is 4 months or less, the violations should be treated as minor.

Table A.5—Continued

Chart C—Non-Minor Misdemeanors: Accessory before or after the fact of a misdemeanor. Assault/Assault and battery. Behind the wheel (regardless of blood alcohol content level). Bigamy. Breaking and entering. Check, worthless, making or uttering, with intent to defraud or deceive ($500 or less). Child neglect. Conspiring to commit misdemeanor. Contributing to delinquency of minor. Criminal mischief. Criminal trespass. Cruelty to animals. Driving while drugged or intoxicated. Failure to stop and render aid after accident. False Imprisonment. Harassment. Indecent exposure. Indecent, insulting, or obscene language communicated directly or by telephone. Juvenile Delinquency involving criminal misconduct. Leaving scene of accident (hit and run). Looting. Motor vehicles: Wrongful appropriation of motor vehicle; joyriding; driving motor vehicle without owner's consent (see Note 2). Negligent homicide. (Contact CNRC, Code 017 for determination.) Prostitution (Contact CNRC, Code 017). Petty larceny (value $500 or less), such as, stealing hubcaps, shoplifting. Possession and/or use of marijuana/controlled substance. (Note 4) Possession of drug paraphernalia. Probation Violation (Contact CNRC Code 017). Providing false information to police/authorities. Reckless driving. (Note 3) Resisting arrest. Sex crime related charges. (Contact CNRC, Code 017 for determination.) Shooting. Slander. Stolen property, knowingly receiving (value $500 or less). Suffrage rights, interference with. Unlawful carrying of firearms; carrying concealed firearm. Unlawful entry. Unlawful use of long-distance telephone lines. Use of telephone to abuse, annoy, harass, threaten, or torment another. Using boat without owner's consent. Willfully discharging firearm so as to endanger life; shooting in public place. Wrongful use of chemical substances. Note 1. The above list is intended as a guide. Offenses of comparable seriousness should be treated as non-minor misdemeanors. In doubtful cases, the following rule should be applied: If the maximum confinement under local law exceeds 4 months but does not exceed one year, the offense should be treated as a non-minor misdemeanor. Note 2. These motor vehicle offenses, and offenses of comparable nature comprise the familiar case of taking or withholding a motor vehicle without authority and with intent to temporarily deprive the owner of his or her property. These are not offenses where the offender intended permanently to deprive the owner of the motor vehicle. Offenses of the latter nature are included in grand larceny or embezzlement involving a value of over $500, listed in Chart D, Felonies. Note 3. May be treated as a Chart B offense if offense did not involve drugs, alcohol, and reckless endangerment, speeding in excess of 15 miles over the posted speed limit, bodily harm to any person (including the driver) or property damage in excess of $500.00. Ensure it is properly documented in the remarks section of DD 1966/3. Note 4. CNRC Code 017 must be contacted in all cases involving possession and/or use of a controlled substance. Drug offenses will be charted in accordance with state guidance and final adjudication if applicable.

Chart D—Felonies: Accessory before or after the fact of a felony. Adultery. Aggravated assault; assault with dangerous weapon; assault, intentionally inflicting great bodily harm; assault with intent to commit felony. Arson. Attempt to commit a felony. Bomb threat. Breaking and entering with intent to commit a felony. Bribery. Burglary. Carnal knowledge of female under 16. Cattle rustling. Car jacking. Check, worthless, making or uttering, with intent to defraud or deceive (over $500). Child abuse. Concealing knowledge of a felony. Conspiring to commit a felony. Criminal libel. Extortion. Forgery; knowingly uttering or passing forged instrument. Graft. Grand larceny; embezzlement (value over $500). Housebreaking. Indecent acts or liberties with child under 16. Indecent assault. Kidnapping; abduction. Mail matters: abstracting, destroying, obstructing, opening, secreting, stealing, or taking. Mails, depositing obscene or indecent matter in. Maiming; disfiguring. Manslaughter. Murder. Pandering. Perjury; subordination of perjury. Possession and/or use of marijuana/controlled substance. (Note 3) Public record: altering, concealing, destroying, mutilating, obliterating, or removing. Rape. Riot. Robbery. Sedition; solicitation to commit sedition. Selling or leasing weapons to minors. Sodomy. Stalking. Stolen property, knowingly receiving (value over $500). Note 1. The above list is intended as a guide. Offenses of comparable seriousness should be treated as felonies. In doubtful cases, the following rule should be applied: if the maximum confinement under local law exceeds one year, the offense should be treated as a felony. An offense, which is classified as a felony by the state in which it was adjudicated, is considered a felony for the purpose of enlistment eligibility determination, regardless of whether it appears on Chart C. Note 2. The CO or XO must personally interview the applicant, verify that he or she meets all the criteria set forth above and sign the waiver document. Note 3. CNRC Code 017 must be contacted in all cases involving possession and/or use of a controlled substance. Drug offenses will be charted in accordance with state guidance and final adjudication if applicable.

Table A.6
Categories of Crimes: Marine Corps

1. Minor Traffic Offenses: Blocking or retarding traffic. Careless driving. Crossing yellow line, driving left of center. Disobeying traffic lights, signs, or signals. Driving on shoulder. Driving uninsured vehicle. Driving with blocked vision. Driving with expired plates or without plates. Driving without license in possession. Driving without registration or with improper registration. Driving wrong way on one-way street. Failure to have vehicle under control. Failure to keep to right or in lane. Failure to signal. Failure to stop for or yield to pedestrian. Failure to yield right-of-way. Faulty equipment (defective exhaust, horn, lights, mirror, muffler, signal device, steering device, tailpipe, windshield wipers, and so forth). Following too closely. Improper backing: backing into intersection or highway; backing on expressway; backing over crosswalk. Improper blowing of horn. Improper parking: restricted area, fire hydrant, double parking, (excluding overtime parking). Improper passing: Passing on right; in no-passing zone; improper lane change; passing stopped school bus with flashing lights; pedestrian in crosswalk. Improper turn. Invalid or unofficial inspection sticker; failure to display inspection sticker. Leaving key in ignition. License plates improperly displayed or not displayed. Operating overloaded vehicle. Racing, drag racing, contest for speed. Speeding. Spinning wheels, improper start. Zigzagging or weaving in traffic. NOTE: Consider offenses of similar nature and traffic offenses treated as minor by local law enforcement agencies as minor traffic offenses.

2. Serious Traffic Offenses: Failure to comply with officer's directions. Reckless driving (Fines $200 or less).

3. Class 1 Minor Non-Traffic Offenses: Curfew violation. Disturbing the peace. Drinking liquor or alcoholic beverages on train, plane, or other conveyance. Drinking in public (non-disorderly). Dumping refuse near highway, littering. Liquor or alcoholic beverages: unlawful possession, consumption in public place, or open container. Loitering. Mischief (painting water towers, graffiti, throwing water-balloons). Purchase, possession, or consumption of alcoholic beverages by minor (underage drinking). Trespass on property (non criminal). Violation of fireworks law. Violation of fish and game laws.

4. Class 2 Minor Non-Traffic Offenses: Abusive language under circumstances to provoke breach of peace. Altered identification (driver's license, birth certificate, and so forth), when intent is to purchase alcoholic beverages. Committing or creating nuisance. Damaging road signs. Disorderly conduct: creating disturbance, boisterous conduct. Failure to appear, failure to comply with a judgment, failure to answer (or disobeying) a summons, or failure to pay a fine. Fighting, participating in a brawl. Illegal betting or gambling: operating illegal handbook, raffle, lottery, punch board, watching cockfight. Juvenile non-criminal misconduct: beyond parental control, incorrigible, runaway, truant, or wayward. Possession of indecent publications or pictures (other than child pornography offenses). Theft, shoplifting (value $100 or less): only if committed under 16 years of age. Unlawful assembly. Vagrancy. Vandalism: injuring or defacing public property or property of another; shooting out street lights; or similar offenses where damage is assessed at $200 or less. NOTE: Consider offenses of a similar nature as minor non-traffic offenses. In doubtful cases, apply the following rule: If the maximum confinement under state or local law is 6-months, or less, treat the offense as a Class 2 minor non-traffic offense.

5. Serious Offenses: Adultery. Assault consummated by battery. Carrying concealed weapon; possession of brass knuckles. Check, worthless, making or uttering, with intent to defraud or deceive ($500 or less). Child pornography offenses. Conspiring to commit misdemeanor. Contempt of court (includes non-payment of child support or alimony required by court order). Contributing to delinquency of minor (includes purchase of alcoholic beverages). Criminal trespass. Desecration of grave. Discharging firearm through carelessness or within municipal limits. Driving while drunk, impaired, intoxicated, or under the influence of alcohol or drugs. Drunk and disorderly and related offenses. Failure to stop and render aid after accident. Fornication. Indecent exposure. Indecent, insulting, or obscene language communicated directly or by telephone. Killing domestic animal. Leaving scene of accident (hit and run) involving no personal injury and property damage is under $500. Liquor or alcoholic: unlawful manufacture or sale. Looting. Malicious/criminal mischief: throwing rocks on highway, throwing missiles at athletic contests, or throwing objects at vehicle. Negligent homicide. Petty larceny; embezzlement (value $500 or less). Possession of marijuana under 30 grams or steroids (requires District waiver). Prostitution/Solicitation. Reckless driving (when fine assessed is $201 or more). Removing property under lien. Removing property from public grounds. Resisting arrest, fleeing and eluding. Selling, leasing, or transferring weapons to minor or unauthorized individual. Slander. Shooting from highway or on public road. Stolen property, knowingly receiving (value $500 or less). Theft, shoplifting (value $500 or less). (If under age 16 and value is $100 or less, treat as class 2 minor non-traffic offense). Unlawful carrying of firearms; carrying concealed firearm. Unlawful entry. Use of telephone to abuse, annoy, harass, threaten, or torment another. Vandalism:

Table A.6—Continued

injuring or defacing public property or property of another; shooting out street lights; or similar offenses where damage is assessed at over $200. Willfully discharging firearm so as to endanger life; shooting in a public place. Wrongful appropriation of motor vehicle; joyriding; driving motor vehicle without owner's consent (if intent is to permanently deprive owner of vehicle, consider as grand larceny under felony offenses below). NOTE: Consider offenses of comparable seriousness as serious offenses. In doubtful cases, apply the following rule: If the maximum confinement under state or local law exceeds 6-months or is equal to or less than 1-year, treat offense as a serious offense.

Felony Offense: Aggravated assault; with dangerous weapon; assault intentionally inflicting great bodily harm; assault with intent to commit felony. Assault and battery on law enforcement officer or child under 16 years of age. Arson. Attempt to commit felony. Breaking and entering (all types). Bribery. Bigamy. Burglary. Carnal knowledge of child under 16. Check, worthless, making or uttering, with intent to defraud or deceive ($501 or more) Conspiring to commit felony. Criminal libel. Draft evasion. Extortion. Forgery; knowingly uttering or passing forged instrument (except for altered identification for purchase of alcoholic beverages). Grand larceny; embezzlement (value $501 or more). Housebreaking. Illegal drugs. Impersonating a police officer, civil official, military officer. Indecent acts or liberties with child under 16, molestation. Indecent assault. Kidnapping, abduction. Leaving scene of accident (hit and run) involving personal injury and/or property damage is over $500. Mail matter: abstracting, destroying, obstructing, opening, secreting, stealing, or taking. Mail, depositing obscene or indecent matter. Maiming; disfiguring. Manslaughter. Murder. Obstructing justice. Pandering. Perjury. Public record; altering, concealing, destroying, mutilating, obliterating, or removing. Rape. Riot. Robbery. Sedition; soliciting to commit sedition. Sodomy. Stolen property, knowingly receiving (value over $500). Theft, shoplifting (value over $500). NOTE: Consider offenses of comparable seriousness as a felony. In doubtful cases, apply the following rule: If maximum confinement under state or local law exceeds 1-year, treat the offense as a felony.

Substance Use

Each service has standards requiring the use of controlled substances.

Army. The Army lists the following as disqualifications that are not waiverable (AR 601-210).

- Alcoholism (p. 28)
- Drug dependence (p. 28)
- Applicants having history of chronic cannabis (marijuana) use or psychological cannabis dependence (p. 29)
- Applicants with three or more convictions or other adverse dispositions for driving while intoxicated, drugged, or impaired in the 5 years preceding application for enlistment (p. 29)
- Subjects of initial civil court conviction or other adverse dispositions for sale, distribution, or trafficking (including "intent to") of cannabis (marijuana), or any other controlled substance (p.29)

Air Force. (See AETC Instruction 36-2002) The Air Force has a more extensive policy than the Army and permanently disqualifies an applicant (waivers are not authorized) if s/he

- self-admits to, is convicted of, or is adversely adjudicated for selling, supplying, or transferring marijuana
- is psychologically dependent or a chronic user or uses marijuana after signing AF Form 2030 (USAF Drug and Alcohol Abuse Certificate)

- self-admits to, is convicted of, or adversely adjudicated for illegal possession, use, sale, or transfer of narcotics, cocaine, lysergic acid diethyl amide (LSD), phencyclidine (PCP "angel dust"), or any other hallucinogen or illegal drug
- is convicted or adversely adjudicated for illegal possession, use, sale, or transfer of amphetamines (including "speed" and synthetics) barbiturates, over-the-counter drugs, or anabolic androgenic steroids.
- is identified during accession drug testing (DAT) as a drug user or has a blood alcohol content of .05 or higher
- self-admits to illegal drug use or involvement by a member of any component of the armed forces while a member (includes Reserve or National Guard) (p. 85).

The Air Force does permit waivers if an applicant:

- self-admits to illegal or wrongful use of amphetamines (includes "speed" and synthetics), barbiturates, over the counter drugs, or anabolic androgenic steroids
- self-admits to or has ever been diagnosed as an alcohol abuser and has abstained for a minimum of two years
- is or was involved in rehabilitation program regarding the use or abuse of marijuana (includes all cannabinoid substances)
- is convicted or adversely adjudicated for possession of drug paraphernalia
- is convicted or adversely adjudicated for illegal possession, use, sale, or transfer of inhalants (glue, paints, thinners, aerodols, amyl/butyl nitrates, and others)
- has reasonable doubt that he or she actually took a disqualifying drug (such as laced in food or other substances) (p. 85).

Navy. (See COMNAVCRUITCOMINST 1130.8F) The Navy does not allow waivers for applicants who have

- been convicted or adversely adjudicated of drug trafficking/supplying
- used LSD within 2 years prior to enlistment (p. 2-121).

However, the Navy does authorize waivers for

- two BTW [Behind the Wheel] offenses.
- prior psychological or physical dependence upon any drug or alcohol
- use of stimulant or depressant drugs, narcotics hallucinogenic or psychedelic drugs
- any drug use while in DEP. (positive non-instrumented drug testing results while in DEP count as drug abuse in DEP.)
- two or more drug paraphernalia related offenses
- two or more drug or alcohol related offenses (p. 2-121).

Marine Corps. (See MCO P1100.72C) The Marine Corps does not grant waivers to applicants who:

- have a history of drug or alcohol dependency or addiction
- have a court conviction for any drug offense (except simple possession of cannabis [30 grams or less], and steroids)
- admit to trafficking marijuana or other illegal drugs (p. 3-66).

The Marine Corps does grant waivers (at a variety of different levels) for applicants who admit to involvement with

- pre-DEP marijuana (1-50 [uses])/steroid abuse or specific prescription drug use
- pre-DEP marijuana (51-200 [uses])/steroid abuse (other than experimentation) or any pre-service drug use if use was over six months prior to DEP
- in-DEP marijuana/steroid use
- in-DEP drug use other than marijuana/steroids
- pre-DEP marijuana (201+ [uses])/ preservice use of cocaine, inhalants, narcotics, opiates, hallucinogens, peyote, or psychoactives if used within six months of DEP-in (p. 3-69).

English Language Training Programs

Army

The Army has three programs that train enlisted recruits in the English language. The oldest of these programs, called the English as a Second Language (ESL) Enlistment Option, was started nearly 20 years ago. This program is designed for recruits who are successfully enlisted into the Army and assigned a Military Operational Specialty (MOS)[3] but require additional English skills before they join their units. A branch of this program, referred to as "09C," has been added to the ESL enlistment option. It is very similar but trains recruits who are not assigned an MOS. These recruits retake the ASVAB after receiving English language training and are then assigned an MOS based on their new AFQT score. The second program, the Foreign Language Recruiting Initiative (FLRI) was started in January of 2001. This program is a pilot that uses a Spanish-language screening test to identify high-aptitude Spanish speakers who score poorly on the AFQT.[4] The program then provides these recruits with English language training, has them retake the AFQT, and assigns MOSs based on the new test score. This program is similar to the 09C branch of the ESL enlistment option except that the first program uses a non-English aptitude test. The third program is called the IRR (Individual Ready Reserve) Direct Arabic Linguist Program, or the 09L program. This program was started on August 11, 2003, and is being used to recruit and train native speakers of languages spoken in Afghanistan and Iraq. Below are more detailed descriptions of each program.

ESL Enlistment Option. This program is tailored generally to Category IIIB recruits[5] who do not score 70 or above on the ECLT. The following instructions are given to recruiters:

---

[3] MOS assignments are directly correlated to AFQT scores.

[4] Applicants must be Category IVA, which is equivalent to the 21st to 30th percentile on the AFQT, in order to qualify.

[5] Those who test in the 31st to 49th percentiles on the AFQT.

Identify all applicants who have difficulty speaking or understanding English (including permanent resident aliens residing in the United States less than 1 year, whose native tongue is not English, and all NPS [non–prior service] applicants from Puerto Rico). (1) Inform identified applicants that they will be taking an English Comprehension Level Test (ECLT) at MEPS. (2) Advise all identified applicants that those who score 69 or below on the ECLT will be required to take English language training prior to IET [Initial Entry Training]. (AR-601-210, p. 35).

Recruits who are required to take English language training are sent to Lackland Air Force base where all the Army English language programs take place.

09C. This program is a derivative of the ESL Enlistment Option program. Recruits in 09C have the opportunity to improve their English language skills and then retake the ASVAB before being assigned an MOS. With improved AFQT scores these recruits are often able to open up a wider range of possible MOSs. This program graduates fewer than 200 graduates per year.

Foreign Language Recruiting Initiative (FLRI). The FLRI is a pilot program that is being run in conjuncture with the Human Resources Research Organization (HumRRO). The following description of the program is found on HumRRO's homepage:[6]

Given the rapid growth of the Hispanic population in the U.S. and the Census Bureau's projections for its continued growth in decades to come, the Army is conducting a pilot test of a Spanish language entrance screening test that may potentially expand the Army's recruiting market. The exam allows the Army to access high aptitude Spanish speakers with limited English ability and then provides English language training to them before they join their unit. This is referred to as the Foreign Language Recruiting Initiative (FLRI). The Army contracted with HumRRO to design, help implement, and evaluate a pilot test of the FLRI concept. HumRRO identified two exams to investigate in this pilot —the Spanish language version of the Wonderlic and the Prueba de Aptitud Academica (PAA). HumRRO has provided support for the FLRI pilot (e.g., providing test materials and instructions to test administrations) and also investigated administrative aspects of the current recruiting and testing processes to design an implementation plan for the pilot FLRI program. We have collected both archival and new data to establish score equating for the Spanish-language tests and the Armed Forces Qualification Test (AFQT) so an appropriate cutoff could be set for the Spanish-language test. Finally, HumRRO designed an evaluation study that will determine the effectiveness of the pilot program. A preliminary evaluation using data collected in the first 10–12 months will involve analyses of Spanish test scores, AFQT scores, English language training performance, and other relevant variables. A final evaluation will be conducted 18-20 months after data collection begins. This evaluation will include analysis of basic training performance outcomes for pilot test participants who have gotten to that stage of their new Army careers.

Although this pilot program tests recruits with a Spanish Language test, recruits must first take the AFQT and score in Category IVA (21st to 30th percentile) to qualify for this program.

---

6 See http://www.humrro.org/corpsite/node/91 (as of September 24, 2008).

For its first two years, this program was run only in Puerto Rico, Miami, New York, Los Angeles, Chicago, and San Antonio. In 2004, it was expanded to Dallas, San Diego, Houston, Sacramento, and Phoenix. However, during the first two years the vast majority of those who qualified and entered the program were from Puerto Rico. Of nearly 400 graduates (there is a cap of 200 recruits per year) more than 90 percent were from Puerto Rico. It is believed that having received some formal education in Spanish greatly increases scores on the Spanish aptitude test. Therefore, some people attribute the large number of Puerto Ricans in the program to the fact that they are the only group that receives their formal education in Spanish.[7]

The test used in this program may also contribute to the high number of Puerto Ricans enrolled. There are different versions of the test—generic Spanish, Mexican Spanish, other dialects, etc.—but the only version currently being used is the generic Spanish. Some Army personnel feel that this test may favor the Spanish spoken by Puerto Ricans and is contributing to the overwhelming number of Puerto Ricans in the program. It is possible that the Mexican version of the test will be available soon, but skeptics argue that there are very small differences in the tests; they do not believe it will significantly impact the enrollment numbers.

IRR (Individual Ready Reserve) Direct Arabic Linguist Program. This program was started on August 11, 2003, because of a demand for native speakers of a variety of languages spoken in Iraq and Afghanistan. The qualifications to enter this program are very limited and recruits are accepted with very low AFQT scores, some below the 10th percentile. Since its inception, 164 recruits have been enlisted in the program, 20 have graduated and 34 are in training. The students come from 24 states, and most are green card holders, although a few citizens have enlisted as well.[8]

Navy

In addition to the Army, the Navy is the only other military service that provides recruits with English language training. However, the Navy's program, called Fundamental Applied Skills Training (FAST), is designed to provide training for a variety of skills, not exclusively English language skills. FAST targets all applicants who score poorly on the Verbal Expression (VE) section of the ASVAB. This includes a wide variety of applicants—some who have difficulties with the English Language and other who have different academic problems. The following instructions are given to recruiters:

> Each applicant entering DEP, who has a VE score of 42 or less, must sign the Fundamental Applied Skills Training (FAST) Administrative Remarks (NAVPERS 1070/613), page 13. . . . Recruiters must brief DEP members on the FAST program. Briefings on FAST should emphasize that FAST offers recruits expert training by college instructors "at no cost" on how to effectively study and learn. Native English speakers complete a 3-week course of instruction: 2 weeks of Navy Reading Skills and one week of Study Skills. Some nonnative English speakers experiencing particular difficulty with English complete a 4-week course of instruction: 3 weeks of Verbal Skills and one week of Study Skills. In very rare instances, a 5th week of instruction for nonnative English speakers is warranted. In the short term, this will maximize their chances of success in classroom training at RTC, and in long

---

7   Telephone conversation with Naomi Verdugo, January 27, 2004.

8   Telephone conversation with Naomi Verdugo, January 27, 2004.

term provides them skills they will use throughout their Navy careers and beyond (COM-NAVCRUITCOMINST 1130.8F, pp. 4-2, 4-3).

An article describing the various programs offered at the Navy's training center described FAST in the following paragraph:

> For recruits with limited literacy or verbal skills, Great Lakes' Fundamental Applied Skills Training (FAST) is available. Recruits with low evaluation test scores, poor study habits, academic performance problems or English language difficulties are assigned here and spend one to five weeks getting their academic background up to a level that will allow them to succeed in the training environment (Wallach, undated).

Although COMNAVCRUITCOMINST 1130.8F states that every applicant "who has a VE score of 42 or less, must sign the Fundamental Applied Skills Training (FAST) Administrative Remarks," not every applicant who signs the Administrative Remarks attends FAST. Because everyone may not perform to his or her true level on the ASVAB, a second test, the Test of Adult Basic Education (TABE), is given to all those who score below a 42. Those who score a 34 or below on the TABE (also equated with an 8th grade level) are admitted to FAST. Approximately 47 percent of those who score below 42 on the VE section of the ASBAV do not actually enter FAST. Of those 47 percent only 1–2 percent are eventually redirected to FAST.[9] When recruits are shipped to FAST, they take an additional test, the ALCPT (American Language Course Placement Test). This test splits the group into those who need ESL training, which is called the Verbal Skills section of FAST, and those who are native English speakers but require literacy training, who are put in the literacy training section of FAST. During FY 02, there were 228 students enrolled in the Verbal Skills section of FAST; of those students, 106 were Hispanic. In FY 03, 173 students were enrolled in the Verbal Skills section; 56 were Hispanic.[10]

The number of recruits sent to FAST is directly proportionate to the number of recruits needed in any given year and thus the standards associated with those needs. Generally, when more recruits are needed, the standards are lower and a higher percentage of recruits (out of a bigger pool of recruits) are sent to FAST. When fewer recruits are needed, the standards rise and fewer are sent. In FY 01, the Navy had a goal of 56,000 recruits and the lowest VE score accepted was 32. Of those enlisted, 5–7 percent attended FAST. In FY 03, the goal was 37,000, and the minimum VE score was 37, so only 3 percent of those enlisted attended FAST. Thus, the population of FAST varies greatly from year to year depending on the Navy's enlistment goals.[11]

---

[9] Telephone conversation with Donald Gunter, January 2004.

[10] Telephone conversation with Donald Gunter, January 2004.

[11] Telephone conversation with Donald Gunter, January 2004.

Homosexual Conduct

DoD Directive 1304.26, which stipulates the requirement regarding homosexual conduct, states the following:

E1.2.8.1. A person's sexual orientation is considered a personal and private matter, and is not a bar to service entry or continued service unless manifested by homosexual conduct in the manner described in paragraph E1.2.8.2, below. Applicants for enlistment, appointment, or induction shall not be asked or required to reveal whether they are heterosexual, homosexual or bisexual. Applicants also will not be asked or required to reveal whether they have engaged in homosexual conduct, unless independent evidence is received indicating that an applicant engaged in such conduct or unless the applicant volunteers a statement that he or she is a homosexual or bisexual, or words to that effect.

E1.2.8.2. Homosexual conduct is grounds for barring entry into the Armed Forces, except as otherwise provided in this section. Homosexual conduct is a homosexual act, a statement by the applicant that demonstrates a propensity or intent to engage in homosexual acts, or a homosexual marriage or attempted marriage. Propensity to engage in homosexual acts means more than an abstract preference or desire to engage in homosexual acts; it indicates a likelihood that a person engages in or will engage in homosexual acts.

E1.2.8.2.1. An applicant shall be rejected for entry into the Armed Forces if, in the course of the accession process, evidence is received demonstrating that the applicant engaged in, attempted to engage in, or solicited another to engage in a homosexual act or acts, unless there is a further determination that:

E1.2.8.2.1.1. Such acts are a departure from the applicant's usual and customary behavior;

E1.2.8.2.1.2. Such acts, under all the circumstances, are unlikely to recur;

E1.2.8.2.1.3. Such acts were not accomplished by use of force, coercion, or intimidation, and;

E1.2.8.2.1.4. The applicant does not have a propensity or intent to engage in homosexual acts.

Such a determination will be made in the course of the normal accession process. A homosexual act means:

E1.2.8.2.1.4.1. Any bodily contact, actively undertaken or passively permitted, between members of the same sex for the purpose of satisfying sexual desires, and

E1.2.8.2.1.4.2. Any bodily contact that a reasonable person would understand to demonstrate a propensity or intent to engage in an act described in subparagraph E1.2.8.2.1.4.1, above.

E1.2.8.2.2. An applicant shall be rejected for entry if he or she makes a statement that he or she is a homosexual or bisexual, or words to that effect, unless there is a further determination that the applicant has demonstrated that he or she is not a person who engages in, attempts to engage in, has a propensity to engage in, or intends to engage in homosexual acts. Such a determination will be made in the course of the normal accession process.

E1.2.8.2.3. An applicant shall be rejected for entry if, in the course of the accession process, evidence is received demonstrating that an applicant has married or attempted to marry a person known to be of the same biological sex (as evidenced by the external anatomy of the persons involved).

E1.2.8.3. Applicants will be informed of separation policy (Section 654 of 10 U.S.C. (reference (a)). Failure to receive such information shall not constitute a defense in any administrative or disciplinary proceeding.

E1.2.8.4. Nothing in these procedures requires rejection for entry into the Armed Forces when the relevant Military Service Command authority determines:

E1.2.8.4.1. That an applicant or inductee made a statement, engaged in acts, or married or attempted to marry a person of the same sex for the purpose of avoiding military service, and

E1.2.8.4.2. Rejection of the applicant or inductee would not be in the best interest of the Armed Forces (p. 9, 10, 11)

Height and Weight Standards

Tables A.7–A.14 provide detailed information on the services' height and weight standards.

Table A.7
Height/Weight Chart: Army Males (Military Acceptable Weight as Related to Age and Height for Males—Initial Army Procurement 1, 2)

Height (inches)	Minimum at Any Age	Weight (lbs)			
		Maximum, by Age			
		17–20	21–27	28–39	40 and Over
60	100	139	141	143	146
61	102	144	146	148	151
62	103	148	150	153	156
63	104	153	155	158	161
64	105	158	160	163	166
65	106	163	165	168	171
66	107	168	170	173	177
67	111	174	176	179	182
68	115	179	181	184	187
69	119	184	186	189	193
70	123	189	192	195	199
71	127	194	197	201	204
72	131	200	203	206	210
73	135	205	208	212	216
74	139	211	214	218	222
75	143	217	220	224	228
76	147	223	226	230	234
77	151	229	232	236	240
78	153	235	238	242	247
79	159	241	244	248	253
80	166	247	250	255	259
Maximum body fat by age		24%	26%	28%	30%

NOTES: (1) If a male exceeds these weights, percentage body fat will be measured by the method described in AR 600–9. (2) If a male also exceeds this body fat percentage, he will be rejected for service.

Table A.8
Height/Weight Chart: Army Females (Military Acceptable Weight as Related to Age and Height For Females—Initial Army Procurement 1, 2)

Height (inches)	Minimum at Any Age	Weight (lbs)			
		Maximum, by Age			
		17–20	21–27	28–39	40 and Over
58	90	112	115	119	122
59	92	116	119	123	126
60	94	120	123	127	130
61	96	124	127	131	135
62	98	129	132	137	139
63	100	133	137	141	144
64	102	137	141	145	148
65	104	141	145	149	153
66	106	146	150	154	158
67	109	149	154	159	162
68	112	154	159	164	167
69	115	158	163	168	172
70	118	163	168	173	177
71	122	167	172	177	182
72	125	172	177	183	188
73	128	177	182	188	193
74	130	183	189	194	198
75	133	188	194	200	204
76	136	194	200	206	209
77	139	199	205	211	215
78	141	204	210	216	220
79	144	209	215	222	226
80	147	214	220	227	232
Maximum body fat by age		30%	32%	34%	36%

NOTES: (1) If a female exceeds these weights, percentage body fat will be measured by the method described in AR 600–9. (2) If a female also exceeds this body fat percentage, she will be rejected for service.

Table A.9
Height/Weight Chart: Air Force Males (Maximum Allowable Weight [MAW] Chart: Men)

Height (inches)	MAW	+ 1/4"	+ 1/2"	+ 3/4"	Desired Weight	Minimum Weight
58	149					98
59	151					99
60	153	153.5	154	154.5	138	100
61	155	155.75	156.5	157.25	140	102
62	158	158.5	159	159.5	142	103
63	160	161	162	163	144	104
64	164	165.25	166.5	167.75	148	105
65	169	170.25	171.5	172.75	152	106
66	174	175.25	176.5	177.75	157	107
67	179	180.25	181.5	182.75	161	111
68	184	185.25	186.5	187.75	166	115
69	189	190.25	191.5	192.75	170	119
70	194	195.25	196.5	197.75	175	123
71	199	200.5	202	203.5	179	127
72	205	206.5	208	209.5	185	131
73	211	212.75	214.5	216.25	190	135
74	218	219.5	221	222.5	196	139
75	224	225.5	227	228.5	202	143
76	230	231.5	233	234.5	207	147
77	236	237.5	239	240.5	212	151
78	242	243.5	245	246.5	218	153
79	248	249.5	251	252.5	223	157
80	254	255.5	257	258.5	229	161

NOTES: For every inch under 60 inches, subtract two pounds from the MAW. For every inch over 80 inches, add six pounds to the MAW. 1. Measure without shoes. 2. Subtract three pounds for clothing. In the Air Force, the MAW chart is used only for weight screening. Individuals who exceed their maximum allowable weight, according to the chart, are given a body-fat measurement to determine if they exceed allowable body-fat standards. Body-fat standards are 20 percent for males 29 years old and younger and 24 percent for males 30 years old and older.

Table A.10
Height/Weight Chart: Air Force Females (Maximum Allowable Weight Chart: Women)

Height (inches)	MAW	+ 1/4"	+ 1/2"	+ 3/4"	Desired Weight	Minimum Weight	+ 1/2"	+ 3/4"	Desired Weight	Minimum Weight
58	132					88				98
59	134					90				99
60	136	136.5	137	137.5	122	92	154	154.5	138	100
61	138	138.75	139.5	140.25	124	95	156.5	157.25	140	102
62	141	141.25	141.5	141.75	127	97	159	159.5	142	103
63	142	143	144	145	128	100	162	163	144	104
64	146	147	148	149	131	103	166.5	167.75	148	105
65	150	151.25	152.5	153.75	135	106	171.5	172.75	152	106
66	155	156	157	158	139	108	176.5	177.75	157	107
67	159	160.25	161.5	162.75	143	111	181.5	182.75	161	111
68	164	165	166	167	148	114	186.5	187.75	166	115
69	168	169.25	170.5	171.75	151	117	191.5	192.75	170	119
70	173	174	175	176	156	119	196.5	197.75	175	123
71	177	178.25	179.5	180.75	159	122	202	203.5	179	127
72	182	183.5	185	186.5	164	125	208	209.5	185	131
73	188	189.5	191	192.5	169	128	214.5	216.25	190	135
74	194	195.25	196.5	197.75	175	130	221	222.5	196	139
75	199	200.5	202	203.5	179	133	227	228.5	202	143
76	205	206.25	207.5	208.75	184	136	233	234.5	207	147
77	210	211.25	212.5	213.75	189	139	239	240.5	212	151
78	215	216.5	218	219.5	193	141	245	246.5	218	153
79	221	222.25	223.5	224.75	199	144	251	252.5	223	157
80	226	227.5	229	230.5	203	147	257	258.5	229	161

NOTES: (1) For every inch under 60 inches, subtract two pounds from the MAW. For every inch over 80 inches, add six pounds to the MAW. (2) Measure without shoes. (3) Subtract three pounds for clothing. In the Air Force, the MAW chart is used only for weight screening. Individuals who exceed their maximum allowable weight, according to the chart, are given a body-fat measurement to determine if they exceed allowable body-fat standards. Body-fat standards are 28 percent for females 29 years old and younger and 32 percent for females 30 years old and older.

Table A.11
Height/Weight Chart: Navy

Men		Applicant's Height (inches)	Women	
Weight (lbs)			Weight (lbs)	
Maximum	Minimum		Maximum	Minimum
131	88	58[a]	131	81
136	89	59[a]	136	83
141	90	60	141	85
145	92	61	145	86
150	93	62	149	88
155	94	63	152	90
160	95	64	156	92
165	95	65	160	94
170	96	66	163	95
175	100	67	167	98
181	104	68	170	101
186	107	69	174	104
191	111	70	177	106
196	114	71	181	110
201	118	72	185	113
206	122	73	189	115
211	125	74	194	119
216	129	75	200	122
221	132	76	205	125
226	136	77	211	129
231	138	78	216	132

[a] Males may enlist at 58" or 59" with approved under-height waivers. Females at 58" or 59" do not require an under-height waiver. Females at 57" or less in height are not eligible to enlist (no waivers).

Table A.12
Height/Weight Charts: Marine Corps Male Standards for
Shipping to Recruit Training

Height (inches)	Maximum Weight	
	5% Over Retention Standards	10% Over Retention Standards
58	139	154
59	143	150
60	148	155
61	153	161
62	158	165
63	163	171
64	168	176
65	174	182
66	179	187
67	185	194
68	190	199
69	196	205
70	202	211
71	207	217
72	213	223
73	219	229
74	225	235
75	231	242
76	237	249
77	244	255
78	250	262
79	256	268
80	263	275

NOTES: Recruits may ship to recruit training if they are 5% or less over retention weight and pass the Initial Strength Test (IST). Applicants who are more than 5% over retention weight and pass the IST require a waiver.

Table A.13
Height/Weight Charts: Marine Corps Male Standards for Enlistment into the Delayed Entry Program

Height (inches)	Minimum Weight	Maximum Weight, by Age		
		16–20	21–30	31–35
58	96	148	153	152
59	98	153	158	157
60	100	158	163	162
61	102	163	168	167
62	103	168	174	173
63	104	174	179	178
64	105	179	185	184
65	106	185	191	190
66	107	191	197	196
67	111	197	203	202
68	115	203	209	208
69	119	209	215	214
70	123	215	222	220
71	127	221	228	227
72	131	227	234	233
73	135	233	241	240
74	139	240	247	246
75	143	246	254	253
76	147	253	261	260
77	151	260	268	266
78	153	267	275	273
79	157	274	282	277
80	160	281	288	285

NOTE: These weight standards apply for contracting only (i.e., enlistment into the DEP). Standards for male applicants entering active duty or initial active duty for training (i.e., "shipping" to recruit training) are provided in the previous chart.

Table A.14
Height/Weight Charts: Marine Corps Female Standards for Shipping to Recruit Training and Enlistment into the Delayed Entry Program

Height (inches)	Minimum Weight	Maximum Weight, by Age		
		16–20	21–30	31–35
58	91	120	123	126
59	94	124	127	130
60	97	128	131	134
61	100	132	135	138
62	104	137	140	143
63	107	141	144	147
64	110	146	149	152
65	114	150	153	156
66	117	155	157	161
67	121	160	163	166
68	125	164	167	170
69	128	169	173	175
70	132	174	177	180
71	136	179	181	185
72	140	184	187	190
73	144	189	192	195
74	148	195	197	201
75	152	200	203	206
76	156	205	208	211
77	160	211	214	217
78	164	216	219	222
79	168	222	225	228
80	173	228	231	234

NOTES: Heights below 58 inches (exact measurement) normally will not be waived. Measurements of one-half inch or more will be rounded-up to the next higher inch (except 57.5); measurements of less that one-half inch will be rounded-down to the next lower inch.

Medical Screening

As discussed in Chapter Two, DoD directs that medical screening occur at the MEPS. Two types of medical conditions disqualify applicants for service: those that are waiverable and those that are not. The medical conditions that disqualify applicants are listed in AR 40-501, and are replicated in the following list.

Complete List of Army Medical Requirements from AR 40-501

2-3. Abdominal organs and gastrointestinal system

The causes for rejection for appointment, enlistment, and induction are an authenticated history of:

    a. Esophagus. Ulceration, varices, fistula, achalasia, or other dismotility disorders; chronic or recurrent esophagitis if confirmed by appropriate x-ray or endoscopic examination (530).

    b. Stomach and duodenum.

        (1) Gastritis. Chronic hypertrophic, or severe (535).

        (2) Active ulcer of the stomach or duodenum confirmed by x-ray or endoscopy (533).

        (3) Congenital abnormalities of the stomach or duodenum causing symptoms or requiring surgical treatment (751), except a history of surgical correction of hypertrophic pyloric stenosis of infancy.

    c. Small and large intestine.

        (1) Inflammatory bowel disease. Regional enteritis (555), ulcerative colitis (556), ulcerative proctitis (556).

        (2) Duodenal diverticula with symptoms or sequelae (hemorrhage, perforation, etc.) (562.02).

        (3) Intestinal malabsorption syndromes, including postsurgical and idiopathic (579).

        (4) Congenital (751). Condition, to include Meckel's diverticulum or functional (564) abnormalities, persisting or symptomatic within the past 2 years.

    d. Gastrointestinal bleeding. History of, unless the cause has been corrected, and is not otherwise disqualifying (578).

    e. Hepato-pancreatic-biliary tract.

        (1) Viral hepatitis (070), or unspecified hepatitis (570), within the preceding 6 months or persistence of symptoms after 6 months, or objective evidence of impairment of liver function, chronic hepatitis, and hepatitis B carriers (070). (Individuals who are known to have tested positive for hepatitis C virus (HCV) infection require confirmatory testing. If positive, individuals should be clinically evaluated for objective evidence of liver function impairment. If evaluation reveals no signs or symptoms of disease, the applicant meets the standards.)

        (2) Cirrhosis (571), hepatic cysts and abscess (572), and sequelae of chronic liver disease (572).

        (3) Cholecystitis, acute or chronic, with or without cholelithiasis (574), and other disorders of the gallbladder including post-cholecystectomy syndrome (575), and biliary system (576). *Note.* Cholecystectomy is not disqualifying 60 days postsurgery (or 30 days post-laproscopic surgery), providing there are no disqualifying residuals from treatment.

        (4) Pancreatitis. Acute (577.0) and chronic (577.1).

    f. Anorectal.

        (1) Anal fissure if persistent, or anal fistula (565).

        (2) Anal or rectal polyp (569.0), prolapse (569.1), stricture (569.2), or incontinence (787.6).

        (3) Hemorrhoids, internal or external, when large, symptomatic, or history of bleeding (455).

    g. Spleen.

        (1) Splenomegaly, if persistent (789.2).

(2) Splenectomy (P41.5), except when accomplished for trauma, or conditions unrelated to the spleen, or for hereditary spherocytosis (282.0).

h. Abdominal wall.

(1) Hernia, including inguinal (550), and other abdominal (553), except for small, asymptomatic umbilical or asymptomatic hiatal.

(2) History of abdominal surgery within the preceding 60 days (P54), except that individuals post-laparoscopic cholecystectomy may be qualified after 30 days.

i. Other.

(1) Gastrointestinal bypass (P43) or stomach stapling (P44) for control of obesity.

(2) Persons with artificial openings (V44).

2-4. Blood and blood-forming tissue diseases

The causes for rejection for appointment, enlistment, and induction are an authenticated history of:

a. Anemia. Any hereditary (282), acquired (283), aplastic (284), or unspecified (285) anemia that has not permanently corrected with therapy.

b. Hemorrhagic disorders. Any congenital (286) or acquired (287) tendency to bleed due to a platelet or coagulation disorder.

c. Leukopenia. Chronic or recurrent (288), based upon available norms for ethnic background.

d. Immunodeficiency (279).

2-5. Dental

The causes for rejection are for appointment, enlistment, and induction are:

a. Diseases of the jaw or associated tissues which are not easily remediable, and will incapacitate the individual or otherwise prevent the satisfactory performance of duty. This includes temporomandibular disorders (524.6) and/or myofascial pain dysfunction that is not easily corrected or has the potential for significant future problems with pain and function.

b. Severe malocclusion (524) that interferes with normal mastication or requires early and protracted treatment; or relationship between mandible and maxilla that prevents satisfactory future prosthodontic replacement.

c. Insufficient natural healthy teeth (521) or lack of a serviceable prosthesis, preventing adequate mastication and incision of a normal diet. This includes complex (multiple fixture) dental implant systems that have associated complications that severely limit assignments and adversely affect performance of world–wide duty. Dental implants systems must be successfully osseointegrated and completed.

d. Orthodontic appliances for continued treatment (V53.4) (attached or removable). Retainer appliances are permissible, provided all active orthodontic treatment has been satisfactorily completed.

2-6. Ears

The causes for rejection for appointment, enlistment, and induction are:

a. External ear. Atresia or severe microtia (744), acquired stenosis (380.5), severe chronic or acute otitis externa (380.2), or severe traumatic deformity (738.7).

b. Mastoids. Mastoiditis (383), residual of mastoid operation with fistula (383.81), or marked external deformity that prevents or interferes with wearing a protective mask or helmet (383.3).

c. Meniere's Syndrome. Or other diseases of the vestibular system (386).

d. Middle and inner ear. Acute or chronic otitis media (382), cholesteatoma (385.3), or history of any inner (P20) or middle (P19) ear surgery excluding myringotomy or successful tympanoplasty.

e. Tympanic membrane. Any perforation of the tympanic membrane (384), or surgery to correct perforation within 120 days of examination (P19).

2-7. Hearing

The cause for rejection for appointment, enlistment, and induction is a hearing threshold level greater than that described in paragraph c below.

a. Audiometers, calibrated to standards of the International Standards Organization (ISO 1964) or the American National Standards Institute (ANSI 1996), will be used to test the hearing of all applicants.

b. All audiometric tracings or audiometric readings recorded on reports of medical examination or other medical records will be clearly identified.

c. Acceptable audiometric hearing levels (both ears) are:

(1) Pure tone at 500, 1000, and 2000 cycles per second of not more than 30 decibels (dB) on the average (each ear), with no individual level greater than 35dB at these frequencies.

(2) Pure tone level not more than 45 dB at 3000 cycles per second each ear, and 55 dB at 4000 cycles per second each ear.

2-8. Endocrine and metabolic disorders

The causes for rejection for appointment, enlistment, and induction are an authenticated history of:

a. Adrenal dysfunction (255) of any degree.

b. Diabetes mellitus (250) of any type.

c. Glycosuria. Persistent, when associated with impaired glucose tolerance (250) or renal tubular defects (271.4).

d. Acromegaly. Gigantism or other disorder of pituitary function (253).

e. Gout (274).

f. Hyperinsulinism (251.1).

g. Hyperparathyroidism (252.0) and hypoparathyroidism (252.1).

h. Thyroid disorders.

(1) Goiter, persistent or untreated (240).

(2) Hypothyroidism, uncontrolled by medication (244).

(3) Cretinism (243).

(4) Hyperthyroidism (242).

(5) Thyroiditis (245).

i. Nutritional deficiency diseases. Such diseases include beriberi (265), pellagra (265.2), and scurvy (267).

j. Other endocrine or metabolic disorders such as cystic fibrosis (277), porphyria (277.1), and amyloidosis (277.3) that obviously prevent satisfactory performance of duty or require frequent or prolonged treatment.

2-9. Upper extremities

(See also para 2–11.) The causes for rejection for appointment, enlistment, and induction are:

  a. Limitation of motion. An individual will be considered unacceptable if the joint ranges of motion are less than the measurements listed below. Methods of measurement appear in TC 8–640.

   (1) Shoulder (726.1):

     (a) Forward elevation to 90 degrees.

     (b) Abduction to 90 degrees.

   (2) Elbow (726.3):

     (a) Flexion to 100 degrees.

     (b) Extension to 15 degrees.

   (3) Wrist (726.4): a total range of 60 degrees (extension plus flexion) or radial and ulnar deviation combined arc 30 degrees.

   (4) Hand (726.4):

     (a) Pronation to 45 degrees.

     (b) Supination to 45 degrees.

   (5) Fingers and thumb (726.4): inability to clench fist, pick up a pin, grasp an object, or touch tips of at least three fingers with thumb.

  b. Hand and fingers.

   (1) Absence of the distal phalanx of either thumb (885).

   (2) Absence of distal and middle phalanx of an index, middle, or ring finger of either hand, irrespective of the absence or loss of little finger (886).

   (3) Absence of more than the distal phalanx of any two of the following fingers: index, middle finger, or ring finger of either hand (886).

   (4) Absence of hand or any portion thereof (887) except for fingers as noted above.

   (5) Polydactyly (755).

   (6) Scars and deformities of the fingers or hand (905.2) that are symptomatic or that impair normal function to such a degree as to interfere with the satisfactory performance of military duty.

   (7) Intrinsic paralysis or weakness, including nerve palsy (354) sufficient to produce physical findings in the hand such as muscle atrophy or weakness.

   (8) Wrist, forearm, elbow, arm, or shoulder. Recovery from disease or injury with residual weakness or symptoms such as to preclude satisfactory performance of duty (905.2), or grip strength of less than 75 percent of predicted normal when injured hand is compared with the normal hand (non-dominant is 80 percent of dominant grip).

2-10. Lower extremities

(See also para 2–11.) The causes for rejection for appointment, enlistment, and induction are:

  a. Limitation of motion. An individual will be considered unacceptable if the joint ranges of motion are less that the measurements listed below. Methods of measurement appear in TC 8–640.

   (1) Hip (due to disease (726.5), injury (905.2)):

     (a) Flexion to 90 degrees.

     (b) No demonstrable flexion contracture.

     (c) Extension to 10 degrees (beyond 0 degrees).

     (d) Abduction to 45 degrees.

(e) Rotation of 60 degrees (internal and external combined).

(2) Knee (due to disease (726.6), injury (905.4)):

(a) Full extension compared with contralateral.

(b) Flexion to 90 degrees.

(3) Ankle (due to disease (726.7), injury (905.4)):

(a) Dorsiflexion to 10 degrees.

(b) Planter flexion to 30 degrees.

(4) Subtalar (due to disease (726.7) or injury (905.4)): eversion and inversion (total to 5 degrees).

b. Foot and ankle.

(1) Absences of one or more small toes (895) if function of the foot is poor or running or jumping is prevented; absence of a foot (896) or any portion thereof except for toes.

(2) Absence of great toe(s) (895); loss of dorsal/plantar flexion if function of the foot is impaired (905.4).

(3) Deformities of the toes, either acquired (735) or congenital (755.66), including poly-dactyly (755.02), that prevent wearing military footwear or impair walking, marching, running, or jumping. This includes hallux valgus (735).

(4) Clubfoot or Pes Cavus (754.5), if stiffness or deformity prevents foot function or wearing military footwear.

(5) Symptomatic pes planus, acquired (734) or congenital (754.6) or pronounced cases, with absence of subtalar motion.

(6) Ingrown toenails (703), if severe.

(7) Planter fascitis (728.7), persistent.

(8) Neuroma (355.6), confirmed condition and refractory to medical treatment or will impair function of the foot.

c. Leg, knee, thigh, and hip.

(1) Loose or foreign bodies within the knee joint (717.6).

(2) Physical findings of an unstable or internally deranged joint (717.9). History of uncorrected anterior (717.83) or posterior (717.84) cruciate ligament injury.

(3) Surgical correction of any knee ligaments if symptomatic or unstable (P81).

(4) History of congenital dislocation of the hip (754.3), osteochondritis of the hip (Legg-Perthes disease) (732.1), or slipped femoral epiphysis of the hip (732.2).

(5) Hip dislocation (835) within 2 years before examination.

(6) Osteochondritis of the tibial tuberosity (Osgood-Schlatter disease) (732.4), if symptomatic.

d. General.

(1) Deformities (905.4), disease or chronic pain (719.4) of one or both lower extremities that have interfered with function to such a degree as to prevent the individual from following a physically active vocation in civilian life or that would interfere with walking, running, or weight bearing, or the satisfactory completion of prescribed training or military duty.

(2) Shortening of a lower extremity (736.81) resulting in a noticeable limp or scoliosis.

2-11. Miscellaneous conditions of the extremities
(See also paras 2-9 and 2-10.) The causes for rejection for appointment, enlistment, and induction are an authenticated history of:

a. Arthritis.

(1) Active, subacute, or chronic arthritis (716).

(2) Chronic osteoarthritis (715.3) or traumatic arthritis (716.1) of isolated joints of more than a minimal degree, which has interfered with the following of a physically active vocation in civilian life or that prevents the satisfactory performance of military duty.

b. Chronic Retro Patellar Knee Pain Syndrome with or without confirmatory arthroscopic evaluation (717.7).

c. Dislocation if unreduced, or recurrent dislocations of any major joint such as shoulder (831), hip (835), elbow (832), or knee (836); or instability of any major joint such as shoulder (718.1), elbow (718.3), or hip (718.5).

d. Fractures.

(1) Malunion or non-union of any fracture (733.8), except ulnar styloid process.

(2) Orthopedic hardware (733.99), including plates, pins, rods, wires, or screws used for fixation and left in place; except that a pin, wire, or screw not subject to easy trauma is not disqualifying.

e. Injury of a bone or joint of more than a minor nature, with or without fracture or dislocation, that occurred within the preceding 6 weeks: upper extremity (923), lower extremity (924), ribs and clavicle (922).

f. Joint replacement (V43.6).

g. Muscular paralysis, contracture, or atrophy (728), if progressive or of sufficient degree to interfere with military service and muscular dystrophies (359).

h. Osteochondritis dessicans (732.7).

i. Osteochondromatosis or Multiple Cartilaginous Exostoses (727.82).

j. Osteoporosis (733).

k. Osteomyelitis (730), active or recurrent.

l. Scars (709.2), extensive, deep, or adherent to the skin and soft tissues that interfere with muscular movements.

m. Implants, silastic or other devices implanted to correct orthopedic abnormalities (V43).

2-12. Eyes

The causes for rejection for appointment, enlistment, and induction are:

a. Lids.

(1) Blepharitis (373), chronic, of more than mild degree.

(2) Blepharospasm (333.81).

(3) Dacryocystitis, acute or chronic (375.3).

(4) Deformity of the lids (374.4), complete or extensive, sufficient to interfere with vision or impair protection of the eye from exposure.

b. Conjunctiva.

(1) Conjunctivitis, chronic (372.1), including trachoma (076) and allergic conjunctivitis (372.13).

(2) Pterygium, (372.4), if encroaching on the cornea in excess of 3 millimeters (mm), interfering with vision, progressive (372.42), or recurring after two operative procedures (372.45).

(3) Xerophthalmia (372.53).

c. Cornea.

(1) Dystrophy, corneal, of any type (371.5), including keratoconus (371.6) of any degree.

(2) Keratorefractive surgery, history of lamellar (P11.7) and/or penetrating kerato-plasty (P11.6). Laser surgery or appliance utilized to reconfigure the cornea is also disqualifying.

(3) Keratitis (370), acute or chronic, which includes recurrent corneal ulcers, erosions (abrasions), or herpetic ulcers (054.42).

(4) Vascularization (370.6) or opacification (371) of the cornea from any cause that is progressive or reduces vision below the standards prescribed in paragraph 2–13 below.

d. Uveitis (364) or iridocyclitis.

e. Retina.

(1) Angiomatosis (759.6), or other congenitohereditary retinal dystrophy (362.7) that impairs visual function.

(2) Chorioretinitis or inflammation of the retina (363), including histoplasmosis, toxo-plasmosis, or vascular conditions of the eye to include Coats' disease, Eales' disease, and retinitis proliferans, unless a single episode of known cause that has healed and does not interfere with vision.

(3) Congenital or degenerative changes of any part of the retina (362).

(4) Detachment of the retina (361), history of surgery for same, or peripheral retinal injury or degeneration that may cause retinal detachment.

f. Optic nerve.

(1) Optic neuritis (377.3), neuroretinitis, secondary optic atrophy, or documented history of attacks of retrobulbar neuritis.

(2) Optic atrophy (377.1), or cortical blindness (377.7).

(3) Papilledema (377.0).

g. Lens.

(1) Aphakia (379.3), lens implant, or dislocation of a lens.

(2) Opacities of the lens (366) that interfere with vision or that are considered to be progressive.

h. Ocular mobility and motility.

(1) Diplopia (386.2), documented, constant or intermittent.

(2) Nystagmus (379.5).

(3) Strabismus (378), uncorrectable by lenses to less than 40 diopters or accompanied by diplopia.

(4) Strabismus, surgery (P15) for the correction of, within the preceding 6 months.

(5) For entrance into the USMA or ROTC programs, the following conditions are also disqualifying: esotropia of over 15 prism diopters; exotropia of over 10 prism diopters; hypertropia of over 5 prism diopters.

i. Miscellaneous defects and conditions.

(1) Abnormal visual fields due to disease of the eye or central nervous system (368.4), or trauma (368.9). Meridianspecific visual field minimums are as follows:

(a) Temporal, 85 degrees.

(b) Superior-temporal, 55 degrees.

(c) Superior, 45 degrees.

(d) Superior nasal, 55 degrees.

(e) Nasal, 60 degrees.

(f) Inferior nasal, 50 degrees.

(g) Inferior, 65 degrees.

(h) Inferior-temporal, 85 degrees.

(2) Absence of an eye, congenital (743) or acquired (360.8).

(3) Asthenopia (368.13), severe.

(4) Exophthalmos (376), unilateral or bilateral, non–familial.

(5) Glaucoma (365), primary, or secondary, or pre-glaucoma as evidenced by intraocular pressure above 21 millimeters of mercury (mmHg), or the secondary changes in the optic disc or visual field loss associated with glaucoma.

(6) Loss of normal pupillary reflex reactions to accommodation (367.5) or light (379.4), including Adie's syndrome.

(7) Night blindness (368.6).

(8) Retained intraocular foreign body (360).

(9) Growth or tumors of the eyelid, other than small basal cell tumors which can be cured by treatment, and small nonprogressive asymptomatic benign lesions.

(10) Any organic disease of the eye (360) or adnexa (376) not specified above, that threatens vision or visual function.

2-13. Vision

The causes for rejection for appointment, enlistment, and induction are:

a. Distant visual acuity of any degree that does not correct with spectacle lenses to at least one of the following (367):

(1) 20/40 in one eye and 20/70 in the other eye.

(2) 20/30 in one eye and 20/100 in the other eye.

(3) 20/20 in one eye and 20/400 in the other eye. However, for entrance into USMA, distant visual acuity that does not correct to 20/20 in each eye is disqualifying. For entrance into ROTC programs and OCS, distant visual acuity that does not correct to 20/20 in one eye and 20/100 in the other eye is disqualifying.

b. Near visual acuity (367) of any degree that does not correct to 20/40 in the better eye.

c. Refractive error (hyperopia (367.0), myopia (367.1), astigmatism (367.2)), in any spherical equivalent of worse than −8.00 or + 8.00 diopters; if ordinary spectacles cause discomfort by reason of ghost images or prismatic displacement; or if corrected by orthokeratology or keratorefractive surgery. However, for entrance into USMA or Army ROTC programs, the following conditions are disqualifying:

(1) Astigmatism, all types over 3 diopters.

(2) Hyperopia over 8.00 diopters spherical equivalent.

(3) Myopia over 6.75 diopters spherical equivalent.

(4) Refractive error corrected by orthokeratology or keratorefractive surgery.

d. Contact lenses. Complicated cases requiring contact lenses for adequate correction of vision, such as corneal scars (371) and irregular astigmatism (367.2).

e. Color vision (368.5). Although there is no standard, color vision will be tested because adequate color vision is a prerequisite for entry into many military specialties. However, for entrance into the USMA or Army ROTC or OCS programs, the inability to distinguish and identify without confusion the color of an object, substance, material, or light that is uniformly colored a vivid red or vivid green is disqualifying.

2–14. Genitalia

The causes for rejection for appointment, enlistment, and induction are:

  a. Female genitalia.

   (1) Abnormal uterine bleeding (626.2), including menorrhagia, metrorrhagia, or polymenorrhea.

   (2) Amenorrhea (626.0), unexplained.

   (3) Dysmenorrhea (625.3), incapacitating to a degree recurrently necessitating absences of more than a few hours from routine activities.

   (4) Endometriosis (617).

   (5) Hermaphroditism (752.7).

   (6) Menopausal syndrome (627), if manifested by more than mild constitutional or mental symptoms, or artificial menopause if less than 1 year's duration.

   (7) Ovarian cysts (620), persistent, clinically significant.

   (8) Pelvic inflammatory disease (614), acute or chronic.

   (9) Pregnancy (V22).

   (10) Uterus, congenital absence of (752.3), or enlargement due to any cause (621.2).

   (11) Vulvar or vaginal ulceration (616.5), including herpes genitalia (054.11) and condyloma acuminatum (078.11), acute or chronic, not amenable to treatment. Such treatment must be given and demonstrated effective prior to accession.

   (12) Abnormal Pap smear (795) graded LGSIL or higher severity, or any smear in which the descriptive terms carcinoma-in-situ, invasive cancer, condyloma acuminatum, human papilloma virus, or dysplasia are used.

   (13) Major abnormalities and defects of the genitalia such as a change of sex (P64.5). A history thereof, or dysfunctional residuals from surgical correction of these conditions.

  b. Male genitalia.

   (1) Absence of both testicles, either congenital (752.8), or acquired (878.2), or unexplained absence of a testicle.

   (2) Epispadias or Hypospadias (752.6), when accompanied by evidence of infection of the urinary tract, or if clothing is soiled when voiding.

   (3) Undiagnosed enlargement or mass of testicle or epididymis (608.9).

   (4) Undescended testicle(s) (752.5).

   (5) Orchitis (604), acute or chronic epididymitis.

   (6) Penis, amputation of (878), if the resulting stump is insufficient to permit normal micturition.

   (7) Penile infectious lesions, including herpes genitalis (054.1) and condyloma acuminata (078.11), acute or chronic, not amenable to treatment. Such treatment must be given and demonstrated effective prior to accession.

   (8) Prostatitis (601), acute or chronic.

   (9) Hydrocele (603.9). Left varicocele, if painful, or any right varicocele (456.4).

  c. Major abnormalities and defects of the genitalia, such as a change of sex (P64.5), a history thereof, or dysfunctional residuals from surgical correction of these conditions.

2-15. Urinary system

(See para 2–8.) The causes for rejection for appointment, enlistment, and induction are:

a. Cystitis (595).

b. Urethritis (597).

c. Enuresis (788.3) or incontinence of urine beyond age 12. (See also para 2–29.)

d. Hematuria, pyuria, or other findings indicative of renal tract disease (599).

e. Urethral stricture (598) or fistula (599.1).

f. Kidney.

    (1) Absence of one kidney, congenital (753.0) or acquired (593.89).

    (2) Infections, acute or chronic (590).

    (3) Polycystic kidney (753.1), confirmed history of.

    (4) Horseshoe kidney (753.3).

    (5) Hydronephrosis (591).

    (6) Nephritis, acute (580) or chronic (582).

g. Proteinuria (791) under normal activity (at least 48 hours after strenuous exercise) greater than 200 milligrams (mg)/24 hours, or a protein to creatinine ratio greater than 0.2 in a random urine sample, unless nephrologic consultation determines the condition to be benign orthostatic proteinuria.

h. Renal calculus (592) within the previous 12 months, recurrent calculus, nephrocalcinosis, or bilateral renal calculi at any time.

2-16. Head

The causes for rejection for appointment, enlistment, and induction are:

a. Injuries, including severe contusions and other wounds of the scalp (920) and cerebral concussion (850), until a period of 3 months has elapsed. (See para 2–26.)

b. Deformities of the skull, face, or jaw (754.0) of a degree that would prevent the individual from wearing a protective mask or military headgear.

c. Defects (756.0), loss or congenital absence of the bony substance of the skull not successfully corrected by reconstructive materials, or leaving residual defect in excess of 1 square inch (6.45 centimeter (cm)2) or the size of a 25 cent piece.

2-17. Neck

The causes for rejection for appointment, enlistment, and induction are:

a. Cervical ribs (756.2), if symptomatic or so obvious that they are found on routine physical examination. (Detection based primarily on x-rays is not considered to meet this criterion.)

b. Congenital cysts (744.4) of branchial cleft origin or those developing from remnants of the thyroglossal duct, with or without fistulous tracts.

c. Contraction (723.8) of the muscles of the neck, spastic or non–spastic, or cicatricial contracture of the neck to the extent that it interferes with wearing a uniform or military equipment or is so disfiguring as to impair military bearing.

2-18. Heart

The causes for rejection for appointment, enlistment, and induction are:

a. All valvular heart diseases, congenital (746) or acquired (394), including those improved by surgery except mitral valve prolapse and bicuspid aortic valve. These latter two conditions are not reasons for rejection unless there is associated tachyarrhythmia, mitral regurgitation, aortic stenosis, insufficiency, or cardiomegaly.

b. Coronary heart disease (410).

c. Symptomatic arrhythmia (or electrocardiographic evidence of arrhythmia), history of.

(1) Supraventricular tachycardia (427.0), or any dysrhythmia originating from the atrium or sinoatrial node, such as atrial flutter, and atrial fibrillation, unless there has been no recurrence during the preceding 2 years while off all medications. Premature atrial or ventricular contractions are disqualifying when sufficiently symptomatic to require treatment or result in physical or psychological impairment.

(2) Ventricularar rhythmias (427.1), including ventricular fibrillation, tachycardia, and multifocal premature ventricular contractions. Occasional asymptomatic premature ventricular contractions are not disqualifying.

(3) Ventricular conduction disorders, left bundle branch block (426.2), Mobitz type II second degree atrioventricular (AV) block (426.12), and third degree AV block (426.0). Wolff-Parkinson-White Syndrome (426.7) and Lown- Ganong-Levine-Syndrome (426.81) associated with an arrhythmia are also disqualifying.

(4) Conduction disturbances such as first degree AV block (426.11), left anterior hemiblock (426.2), right bundle branch block (426.4), or Mobitz type I second degree AV block (426.13) are disqualifying when symptomatic or associated with underlying cardiovascular disease.

d. Hypertrophy or dilatation of the heart (429.3).

e. Cardiomyopathy (425), including myocarditis (422), or history of congestive heart failure (428) even though currently compensated.

f. Pericarditis (420).

g. Persistent tachycardia (785) (resting pulse rate of 100 or greater).

h. Congenital anomalies of heart and great vessels (746), except for corrected patent ductus arteriosus.

2-19. Vascular system

The causes for rejection for appointment, enlistment, and induction are:

a. Abnormalities of the arteries and blood vessels (447), including aneurysms (442), even if repaired, atherosclerosis (440), or arteritis (446).

b. Hypertensive vascular disease (401), evidenced by the average of three consecutive diastolic blood pressure measurements greater than 90 mmHg or three consecutive systolic pressure measurements greater than 140 mmHg.

High blood pressure requiring medication or a history of treatment including dietary restriction.

c. Pulmonary (415) or systemic embolization (444).

d. Peripheral vascular disease, including Raynaud's phenomenon (443).

e. Vein diseases, recurrent thrombophlebitis (451), thrombophlebitis during the preceding year, or any evidence of venous incompetence, such as large or symptomatic varicose veins, edema, or skin ulceration (454).

2-20. Height

The causes for rejection for appointment, enlistment, and induction are:

a. Men: Height below 60 inches or over 80 inches.

b. Women: Height below 58 inches or over 80 inches.

2-21. Weight

   a. Army applicants for initial appointment as commissioned officers (to include appointment as commissioned warrant officers) must meet the standards of AR 600–9. Body fat composition is used as the final determinant in evaluating an applicant's acceptability when the weight exceeds the weight tables.

   b. All other applicants must meet the standards of tables 2-1 and 2-2. Body fat composition is used as the final determinant in evaluating an applicant's acceptability when the weight exceeds the weight tables.

2-22. Body build

The cause for rejection for appointment, enlistment, and induction is deficient muscular development that would interfere with the completion of required training.

2-23. Lungs, chest wall, pleura, and mediastinum

The causes for rejection for appointment, enlistment, and induction are:

   a. Abnormal elevation of the diaphragm (793.2), either side.

   b. Abscess of the lung (513).

   c. Acute infectious processes of the lung (518), until cured.

   d. Asthma (493), including reactive airway disease, exercise induced bronchospasm or asthmatic bronchitis, reliably diagnosed at any age. Reliable diagnostic criteria should consist of any of the following elements:

   (1) Substantiated history of cough, wheeze, and/or dyspnea that persists or recurs over a prolonged period of time, generally more than 6 months.

   (2) If the diagnosis of asthma is in doubt, a test for reversible airflow obstruction (greater than a 15 percent increase in forced expiratory volume in 1 second (FEVI) following administration of an inhaled bronchodilator) or airway hyperactivity (exaggerated decrease in airflow induced by standard bronchoprovocation challenge such as methacholine inhalation or a demonstration of exercise-induced bronchospasm) must be performed.

   e. Bronchitis (490), chronic, symptoms over 3 months occurring at least twice a year.

   f. Bronchiectasis (494).

   g. Bronchopleural fistula (510).

   h. Bullous or generalized pulmonary emphysema (492).

   i. Chronic mycotic diseases (117) of the lung including coccidioidomycosis.

   j. Chest wall malformation (754) or fracture (807) that interferes with vigorous physical exertion.

   k. Empyema (510), including residual pleural effusion (511.9) or unhealed sinuses of chest wall (510).

   l. Extensive pulmonary fibrosis (515).

   m. Foreign body in lung, trachea, or bronchus (934).

   n. Lobectomy, with residual pulmonary disease or removal of more than one lobe (P32.4).

   o. Pleurisy with effusion (511.9), within the previous 2 years if known or unknown origin.

   p. Pneumothorax (512) during the year preceding examination if due to a simple trauma or surgery; during the 3 years preceding examination from spontaneous origin. Recurrent spontaneous pneumothorax after surgical correction or pleural sclerosis.

   q. Sarcoidosis (135). (See para 2–34.)

r. Silicone breast implants, encapsulated (85.53) if less than 9 months since surgery or with symptomatic complications.

s. Tuberculous lesions. (See para 2-34.)

2-24. Mouth

The causes for rejection for appointment, enlistment, and induction are:

a. Cleft lip or palate defects (749), unless satisfactorily repaired by surgery.

b. Leukoplakia (528.6).

2-25. Nose, sinuses, and larynx

The causes for rejection for appointment, enlistment, and induction are:

a. Allergic manifestations.

(1) Allergic or vasomotor rhinitis (477), if moderate or severe and not controlled by oral medications, desensitization, or topical corticosteroid medication.

(2) Atrophic rhinitis (472).

(3) Vocal cord paralysis (478.3), or symptomatic disease of the larynx (478.7).

b. Anosmia or parosmia (352).

c. Epistaxis (784.7), recurrent.

d. Nasal polyps (471), unless surgery was performed at least 1 year before examination.

e. Perforation of nasal septum (478.1), if symptomatic or progressive.

f. Sinusitis (461), acute.

g. Sinusitis, chronic (473), when evidenced by chronic purulent nasal discharge, hyperplastic changes of the nasal tissue, symptoms requiring frequent medical attention, or x–ray findings.

h. Larynx ulceration, polyps, granulated tissue, or chronic laryngitis (476).

i. Tracheostomy (V44) or tracheal fistula.

j. Deformities or conditions (750.9) of the mouth, tongue, palate throat, pharynx, larynx, and nose that interfere with chewing, swallowing, speech, or breathing.

k. Pharyngitis (462) and nasopharyngitis (472.2), chronic.

2-26. Neurological disorders

The causes for rejection for appointment, enlistment, and induction are:

a. Cerebrovascular conditions, any history of subarachnoid (430) or intracerebral (431) hemorrhage, vascular insufficiency, aneurysm, or arteriovenous malformation (437).

b. Congenital malformations (742), if associated with neurological manifestations or if known to be progressive; meningocele (741), even if uncomplicated.

c. Degenerative and hereditodegenerative disorders affecting the cerebrum (330), basal ganglia (333), cerebellum (334), spinal cord (335), and peripheral nerves, or muscles (337).

d. Recurrent headaches (784) of all types if they are of sufficient severity or frequency to interfere with normal function within 3 years.

e. Head injury (854).

(1) Applicants with a history of head injury with—

(a) Late post-traumatic epilepsy (occurring more than 1 week after injury).

(b) Permanent motor or sensory deficits.

(c) Impairment of intellectual function.

(d) Alteration of personality.

(e) Central nervous system shunt.

(2) Applicants with a history of severe head injury are unfit for a period of at least 5 years, after which they may be considered fit if complete neurological and neurophysical evaluation shows no residual dysfunction or complications. Applicants with a history of severe penetrating head injury are unfit for a period of at least 10 years after the injury. After 10 years they may be considered fit if complete neurological and neuropsychological evaluation shows no residuals dysfunction or complications. Severe head injuries are defined by one or more of the following:

(a) Unconsciousness or amnesia, alone or in combination, of 24 hours duration or longer.

(b) Depressed skull fracture.

(c) Laceration or contusion of dura or brain.

(d) Epidural, subdural, subarachnoid, or intracerebral hematoma.

(e) Associated abscess or meningitis.

(f) Cerebrospinal fluid rhinorrhea or otorrhea persisting more than 7 days.

(g) Focal neurologic signs.

(h) Radiographic evidence of retained metallic or bony fragments.

(i) Leptomeningeal cysts or arteriovenous fistula.

(j) Early post-traumatic seizure(s) occurring within 1 week of injury but more than 30 minutes after injury.

(3) Applicants with a history of moderate head injury are unfit for a period of at least 2 years after injury, after which they may be considered fit if complete neurological evaluation shows no residual dysfunction or complications. Moderate head injuries are defined by unconsciousness or amnesia, alone or in combination of 1 to 24 hours duration or linear skull fracture.

(4) Applicants with a history of mild head injury, as defined by a period of unconsciousness or amnesia, alone or in combination, of 1 hour or less, are unfit for at least 1 month after injury; after which they may be acceptable if neurological evaluation shows no residual dysfunction or complications.

(5) Persistent post-traumatic sequelae, as manifested by headache, vomiting, disorientation, spatial disequilibrium, personality changes, impaired memory, poor mental concentration, shortened attention span, dizziness, altered sleep patterns, or any findings consistent with organic brain syndrome are disqualifying until full recovery has been confirmed by complete neurological and neuropsychological evaluation.

f. Infectious diseases.

(1) Meningitis (322), encephalitis (323), or poliomyelitis (045) within 1 year before examination, or if there are residual neurological defects.

(2) Neurosyphilis (094) of any form, general paresis, tabes dorsalis meningovascular syphilis.

g. Narcolepsy (347), sleep apnea syndrome (780.57).

h. Paralysis, weakness, lack of coordination, pain, sensory disturbance (344).

i. Epilepsy (345), beyond the age of 5 unless the applicant has been free of seizures for a period of 5 years while taking no medication for seizure control, and has a normal electro-encephalogram (EEG). All such applicants will have a current neurology consultation with current EEG results. EEG may be requested by the reviewing authority.

j. Chronic disorders such as myasthenia gravis (358) and multiple sclerosis (340).

k. Central nervous system shunts of all kinds (V45.2).

2-27. Disorders with psychotic features

The causes for rejection for appointment, enlistment, and induction are disorders with psychotic features (295).

2-28. Neurotic, anxiety, mood, somatoform, dissociative, or factitious disorders

The causes for rejection for appointment, enlistment, and induction are a history of such disorders (300) resulting in any or all of the below:

a. Admission to a hospital or residential facility.

b. Care by a physician or other mental health professional for more than 6 months.

c. Symptoms or behavior of a repeated nature that impaired social, school, or work efficiency.

2-29. Personality, conduct, and behavior disorders

The causes for rejection for appointment, enlistment, and induction are:

a. Personality (301), conduct (312), or behavior disorders (313) as evidenced by frequent encounters with law enforcement agencies, antisocial attitudes or behavior, which, while not sufficient cause for administrative rejection, are tangible evidence of impaired capacity to adapt to military service.

b. Personality (301), conduct (312), or behavior (313) disorders where it is evident by history, interview, or psychological testing that the degree of immaturity, instability, personality inadequacy, impulsiveness, or dependency will seriously interfere with adjustment in the Army as demonstrated by repeated inability to maintain reasonable adjustment in school, with employers and fellow workers, and with other social groups.

c. Other behavior disorders including but not limited to conditions such as authenticated evidence of functional enuresis (307.6) or encopresis (307.7), sleepwalking (307.6), or eating disorders that are habitual or persistent (307.1 or 307.5) occurring beyond age 12, or stammering (307.0) of such a degree that the individual is normally unable to express himself or herself clearly or to repeat commands.

d. Specific academic skills defects, chronic history of academic skills (314) or perceptual defects (315), secondary to organic or functional mental disorders that interfere with work or school after age 12. Current use of medication to improve or maintain academic skills.

e. Suicide, history of attempted or suicidal behavior (300.9).

2-30. Psychosexual conditions

The causes for rejection for appointment, enlistment, and induction are transsexualism, exhibitionism, transvestitism, voyeurism, and other paraphilias (302).

2-31. Substance misuse

The causes for rejection for appointment, enlistment, and induction are:

a. Alcohol dependence (303).

b. Drug dependence (304).

c. Non-dependent use of drugs characterized by—

(1) The evidence of use of any controlled hallucinogenic, or other intoxicating substance at time of examination (305), when the use cannot be accounted for as the result of a prescription of a physician.

(2) Documented misuse or abuse of any controlled substance (including cannabinoids or anabolic steroids) requiring professional care (305).

(3) The repeated self-procurement and self-administration of any drug or chemical substance, including cannabinoids or anabolic steroids, with such frequency that it appears that the applicant has accepted the use of or reliance on these substances as part of his or her pattern of behavior (305).

d. The use of LSD (305.3) within a 2-year period of the examination.

e. Alcohol abuse (305), use of alcoholic beverages that leads to misconduct, unacceptable social behavior, poor work or academic performance, impaired physical or mental health, lack of financial responsibility, or a disrupted personal relationship.

2-32. Skin and cellular tissues

The causes for rejection for appointment, enlistment, and induction are:

a. Acne (706), severe, or when extensive involvement of the neck, shoulders, chest, or back would be aggravated by or interfere with the wearing of military equipment, and would not be amenable to treatment. Patients under treatment with isotretinoin (Accutane) are medically unacceptable until 8 weeks after completion of course of therapy.

b. Atopic dermatitis (691) or eczema (692), with active or residual lesions in characteristic areas (face, neck, antecubital, and or/popliteal fossae, occasionally wrists and hands), or documented history thereof after the age of 8.

c. Contact dermatitis (692.4), especially involving rubber or other materials used in any type of required protective equipment.

d. Cysts.

(1) Cysts (706.2), other than pilonidal, of such a size or location as to interfere with the normal wearing of military equipment.

(2) Pilonidal cysts (685), if evidenced by the presence of a tumor mass or a discharging sinus. History of pilonidal cystectomy within 6 months before examination is disqualifying.

e. Dermatitis factitia (698.4).

f. Bullous dermatoses (694), such as Dermatitis Herpetiformis, pemphigus, and epidermolysis bullosa.

g. Chronic Lymphedema (457).

h. Fungus infections (117), systemic or superficial types, if extensive and not amenable to treatment.

i. Furunculosis (680), extensive recurrent, or chronic.

j. Hyperhidrosis of hands or feet (780.8), chronic or severe.

k. Ichthyosis, or other congenital (757) or acquired (216) anomalies of the skin such as nevi or vascular tumors that interfere with function or are exposed to constant irritation.

l. Keloid formation (701.4), if the tendency is marked or interferes with the wearing of military equipment.

m. Leprosy (030.9), any type.

n. Lichen planus (697.0).

o. Neurofibromatosis (von Recklinghausen's disease) (237.7).

p. Photosensitivity (692.72), any primary sun-sensitive condition, such as polymorphous light eruption or solar urticaria; any dermatosis aggravated by sunlight such as lupus erythematosus.

q. Psoriasis (696.1), unless mild by degree, not involving nail pitting, and not interfering with wearing military equipment or clothing.

r. Radiodermatitis (692.82).

s. Scars (709.2) that are so extensive, deep, or adherent that they may interfere with the wearing of military clothing or equipment, exhibit a tendency to ulcerate, or interfere with function. Includes scars at skin graft donor or recipient sites if the area is susceptible to trauma.

t. Scleroderma (710.1).

u. Tattoos (709.9) that will significantly limit effective performance of military service or that are otherwise prohibited under AR 670–1.

v. Urticaria (708.8), chronic.

w. Warts, plantar (078.19), symptomatic.

x. Xanthoma (272.2), if disabling or accompanied by hyperlipemia.

y. Any other chronic skin disorder of a degree or nature, such as Dysplastic Nevi Syndrome (448.1), which requires frequent outpatient treatment or hospitalization, or interferes with the satisfactory performance of duty.

2-33. Spine and sacroiliac joints

(See also para 2-11.) The causes for rejection for appointment, enlistment, and induction are:

a. Arthritis (720). (See para 2–11*a.*)

b. Complaint of a disease or injury of the spine or sacroiliac joints with or without objective signs that has prevented the individual from successfully following a physically active vocation in civilian life (724) or that is associated with pain referred to the lower extremities, muscular spasm, postural deformities, or limitation of motion.

c. Deviation or curvature of spine (737) from normal alignment, structure, or function if—

   (1) It prevents the individual from following a physically active vocation in civilian life.

   (2) It interferes with wearing a uniform or military equipment.

   (3) It is symptomatic and associated with positive physical finding(s) and demonstrable by x-ray.

   (4) There is lumbar scoliosis greater than 20 degrees, thoracic scoliosis greater than 30 degrees, and kyphosis or lordosis greater than 55 degrees when measured by the Cobb method.

d. Fusion, congenital (756.15), involving more than two vertebrae. Any surgical fusion (81.0P) is disqualifying.

e. Healed fractures or dislocations of the vertebrae (805). A compression fracture, involving less than 25 percent of a single vertebra is not disqualifying if the injury occurred more than 1 year before examination and the applicant is asymptomatic. A history of fractures of the transverse or spinous processes is not disqualifying if the applicant is asymptomatic.

f. Juvenile epiphysitis (732.6) with any degree of residual change indicated by x-ray or kyphosis.

g. Ruptured nucleus pulposus (722), herniation of intervertebral disk or history of operation for this condition.

h. Spina bifida (741) when symptomatic or if there is more than one vertebra involved, dimpling of the overlying skin, or a history of surgical repair.

i. Spondylolysis (756.1) and spondylolisthesis (738.4).

j. Weak or painful back (724) requiring external support such as a corset or brace; recurrent sprains or strainsrequiring limitation of physical activity or frequent treatment.

2-34. Systemic diseases

The causes for rejection for appointment, enlistment, and induction are:

a. Amyloidosis (277.3).

b. Ankylosing spondylitis (720).

c. Eosinophilic granuloma (277.8) when occurring as a single localized bony lesion and not associated with soft tissue or other involvement should not be a cause for rejection once healing has occurred. All other forms of the Histiocytosis X spectrum should be rejected.

d. Lupus erythematosus (710) and mixed connective tissue disease.

e. Polymyositis/dermatomyositis complex (710).

f. Progressive Systemic Sclerosis (710), including CRST (calcinosis, Raynaud's phenomenon, sclerodactyly, and telangiectasis) variant. A single plaque of localized scleroderma (morphea) that has been stable for at least 2 years is not disqualifying.

g. Reiter's Disease (099.3).

h. Rheumatoid arthritis (714).

i. Rhabdomyolysis (728.9).

j. Sarcoidosis (135), unless there is substantiated evidence of a complete spontaneous remission of at least 2 years duration.

k. Sjogren's Syndrome (710.2).

l. Tuberculosis (010).

    (1) Active tuberculosis in any form or location, or history of active tuberculosis within the previous 2 years.

    (2) One or more reactivations.

    (3) Residual physical or mental defects from past tuberculosis that would preclude the satisfactory performance of duty.

    (4) Individuals with a past history of active tuberculosis MORE than 2 years prior to enlistment, induction and appointment are QUALIFIED IF they have received a complete course of standard chemotherapy for tuberculosis. In addition, individuals with a tuberculin reaction 10 mm or greater and without evidence of residual disease are qualified once they have been treated with chemoprophylaxis.

    (5) Vasculitis (446) such as Bechet's, Wegener's granulomatosis, polyarteritis nodosa.

2-35. General and miscellaneous conditions and defects

The causes for rejection for appointment, enlistment, and induction are:

a. Allergic manifestations (995.0). A reliable history of anaphylaxis to stinging insects. Reliable history of a moderate to severe reaction to common foods, spices, or food additives.

b. Any acute pathological condition, including acute communicable diseases, until recovery has occurred without sequelae.

c. Chronic metallic poisoning with lead, arsenic, or silver (985), or beryllium or manganese (985).

d. Cold injury (991), residuals of, such as: frostbite, chilblain, immersion foot, trench foot, deep–seated ache, paresthesia, hyperhidrosis, easily traumatized skin, cyanosis, amputation of any digit, or ankylosis.

e. Cold urticaria (708.2) and angioedema, hereditary angioedema (277.6).

f. Filariasis (125), trypanosomiasis (086), schistosomiasis (120), uncinariasis (126.9), or other parasitic conditions, if symptomatic or carrier states.

g. Heat pyrexia, heatstroke, or sunstroke (992). Documented evidence of a predisposition (including disorders of sweat mechanism and a previous serious episode), recurrent episodes requiring medical attention, or residual injury (especially cardiac, cerebral, hepatic, and renal); malignant hyperthermia (995.89).

h. Industrial solvent and other chemical intoxication (982).

i. Motion sickness (994.6). An authenticated history of frequent incapacitating motion sickness after the 12th birthday.

j. Mycotic (114) infection of internal organs.

k. Organ transplant recipient (V42).

l. Presence of human immunodeficiency virus (HIV–I) or antibody (042). Presence is confirmed by repeatedly reactive enzyme-linked immunoassay serological test and positive immunoelectrophoresis (Western Blot) test, or other DOD-approved confirmatory test.

m. Reactive tests for syphilis (093) such as the rapid plasma reagin (RPR) test or venereal disease research laboratory (VDRL) followed by a reactive, confirmatory Fluorescent Treponemal Antibody Absorption (FTA–ABS) test unless there is a documented history of adequately treated syphilis. In the absence of clinical findings, the presence of reactive RPR or VDRL followed by a negative FTA–ABS test is not disqualifying if a cause for the false positive reaction can be identified and is not otherwise disqualifying.

n. Residual of tropical fevers, such as malaria (084) and various parasitic or protozoal infestations that prevent the satisfactory performance of military duty.

o. Rheumatic fever (390) during the previous 2 years, or any history of recurrent attacks; Sydenham's chorea at any age.

p. Sleep apnea (780.57).

2-36. Tumors and malignant diseases

The causes for rejection for appointment, enlistment, and induction are:

a. Benign tumors (M8000) that interfere with function, prevent wearing the uniform or protective equipment, would require frequent specialized attention, or have a high malignant potential.

b. Malignant tumors (V10), exception for basal cell carcinoma, removed with no residual. In addition, the following cases should be qualified if on careful review they meet the following criteria: individuals who have a history of childhood cancer who have not received any surgical or medical cancer therapy for 5 years and are free of cancer; individuals with a history of Wilm's tumor and germ cell tumors of the testis treated surgically and/or with chemotherapy after a 2-year disease-free interval off all treatment; individuals with a history of Hodgkin's disease treated with radiation therapy and/or chemotherapy and disease free off treatment for 5 years; individuals with a history of large cell lymphoma after a 2-year disease-free interval off all therapy.

2-37. Miscellaneous

Any condition that in the opinion of the examining medical officer will significantly interfere with the successful performance of military duty or training (796) may be a cause for rejection for appointment, enlistment, and induction.

Service Waiver Policy

This appendix highlights the Army, Air Force, Navy, and Marine Corps policies regarding the use of waivers of enlistment standards. It is organized in four sections, one for each service. The material is quoted from the documents listed in Table 2.1.

Army

The following is from AR 601-210.

> Applicants who do not meet established enlistment standards are not eligible for enlistment unless a waiver is authorized. Commanders cited in this regulation have the authority to approve waivers as appropriate. The burden is on the applicant to prove to waiver authorities that they have overcome their disqualifications for enlistment, and that their acceptance would be in the best interests of the Army. Waiver authorities will consider the "whole person" concept when considering waiver applications.
>
> 4–3. Submission of requests
>
> *a*. Unless indicated other wise in this regulation, requests for waiver and other actions that require approval by the CG, PERSCOM (for RA), will be forwarded to Commander, PERSCOM. If approval is required by the CG, ARPERCEN (for US AR), they will be forwarded to Commander, ARPERCEN, ATTN: DARP-PAT-I, 9700 Page Boulevard, St. Louis, MO 63132-5200. Waivers for multiple administrative disqualifications (for example, misdemeanor and RE Code waivers) will be forwarded to Commander, PERSCOM, for action. Coordination with CG, USAREC, in regard to the medical portion will be made by CG, PERSCOM. [Each type of waiver, i.e. a medical waiver, or a moral waiver, or a dependency status waiver are specifically designated to the CT PERSCOM or the CG ARPERCEN]
>
> *b*. Waivers approved by CG, PERSCOM, for enlistment in the RA may be used for enlistment in the USAR, provided the individual is otherwise qualified. However, waivers approved by CG, ARPERCEN, may not be used for enlistment into the RA.
>
> *c*. When processing moral waivers, the most serious of all offenses arising out of a single act, which results in a civil court conviction or other adverse disposition will be the offense considered for enlistment eligibility purposes. All offenses charged must be revealed, and required documents obtained. However, as an example, a person caught by police during an attempted shoplifting who then resists arrest has committed two separate acts of mis-

conduct. Both must be considered for waiver purposes. Doubtful cases will be referred to HW, USAREC, ATTN: RCRO-PP-WD, Fort Knox, KY 40121-2726.

Air Force

The following is from AETC Instruction 36-2002.

CRITERIA WAIVERS AND ELIGIBILITY DETERMINATIONS

4.1. General Conditions. . . . If an applicant is not qualified, a recruiter may request a criteria waiver under some conditions. When there are questions about a qualified applicant's suitability, the commander must make an eligibility determination. As with waivers and eligibility determinations, forward exceptions to policy or operational standards through the appropriate chain of command.

4.2. Criteria Waivers. Recruiters may initiate moral, dependency, or drug waiver requests on unqualified applicants providing they are, other than the waiverable conditions, highly qualified and motivated to join the Air Force. Applicants must have displayed sufficient mitigating circumstances that clearly justify a waiver. Make all waiver actions in the best interest of the Air Force.

4.2.1. Procedures. Do not schedule applicants under waiver consideration (other than medical) for a physical examination until the waiver is approved. Submit all waiver requests for an individual at one time. For HQ AFRS and group-level waivers, retain originals at the squadron. The waiver authority will enter disapprovals in the fraud file (see paragraph 4.16).

4.2.2. Validity and Waiting Periods. Waivers are valid as follows: NPS–for as long as the applicant is job-committed; HP–12 months after approval; and OTS–6 months and or two selection boards. OTS and HP waivers remain valid until EAD for selects. Disapproved waivers may be resubmitted 6 months after disapproval. If significant mitigating data were not considered in the original request, commanders may submit the waiver with a request for exception to the 6-month period through the group to HQ AFRS/RSOP/RSOC for NPS, OTS, and HP. If approved, HQ AFRS/RSOPA will remove the applicant from the fraud file and return the waiver to the unit for NPS programs and to HQ AFRS/RSOC/RSOH for OTS or HP.

4.3. Eligibility Determinations. An eligibility determination is the formal process by which commanders review circumstances that place doubt on a qualified applicant's suitability. Three outcomes are possible–an applicant is found to be eligible, ineligible, or not qualified. If an applicant is determined not qualified, the case may be pursued through a criteria waiver request.

4.3.1. Procedures. Eligibility determination is a judgment call. As such, documents submitted must provide the facts and background information necessary for commanders to make informed decisions. It is good practice to include other documents that reflect the applicant's moral history. Commanders will document determinations in the remarks section of DD Form 1966, page 3, for NPS and OTS applicants and the addendum to AF Form 24, Application for Appointment as Reserves of the Air Force or USAF Without Component, for HP applicants. Unfavorable determinations will be entered in the fraud file. Do not schedule applicants requiring an eligibility determination for a physical examination until a favorable determination is rendered.

4.3.2. Validity Period. Favorable eligibility determinations remain valid for 12 months from the date the determination is made. Unfavorable determinations are normally permanent; however, an applicant can request to be reconsidered at any time if significant mitigating information was not previously considered.

4.4. Delegation. Only commanders, vice commanders, and deputy commanders can render waivers and eligibility determinations. Commanders at any level may disapprove the request. If the original commander, vice commander, or deputy commander at approval level is not available, elevate the request to the next higher level of command. Approval authority cannot be delegated to a lower level.

Navy

The following is from COMNAVCRUITCOMINST 1130.8F.

. . . b. An applicant who requires a waiver of any enlistment eligibility requirement must not be processed unless they are considered to be a particularly desirable applicant. **Note: Applicants with preservice moral waivers (drug, alcohol, or criminal) are disqualified from overseas assignment for their first duty station.**

. . .

2B-2 District Level Waiver Procedures.

a. "By Direction" Waiver Approval and Documentation Authority.

(1) District Commanding Officers may authorize their Executive Officer (XO), Enlisted Programs Officer (EPO), or Director (DIR) of the Navy Recruiting Processing Station (NAVCRUITPROSTA), to grant "By direction" enlistment and program eligibility determinations that are within the Commanding Officer's authority to grant. Additionally, Commanding Officers may authorize their Executive Officers authority to grant CO level waivers at any time. This authority must be detailed in a "By direction" letter

. . .

(3) RTC Commanding Officer. The Navy Recruit Training Command (RTC) Commanding Officer is granted commensurate waiver authority with NRD Commanding Officers. The following exceptions apply:

(a) Two alcohol/drug related "behind the wheel" (BTW) convictions.

(b) Two convictions of possession/use of marijuana.

(c) Negligent homicide.

(d) Indecent exposure.

(e) Indecent, insulting or obscene language communicated directly or by telephone.

(f) Sex crime related charges or child-molesting.

2B General Waiver Information and Standards for Waiver Consideration

In all instances where this manual requires preaccession waiver authority at the CNRC level, the RTC Commanding Officer will refer post-accession waiver requests to PERS-83.

b. In all cases when a district waiver is being considered:

(1) The specific merits or liabilities of a request are carefully weighed. The "Whole Person" concept is the general rule followed in deliberations, however specific information may take precedence in some cases depending on the nature of the request. An important factor in all deliberations is the relative competitiveness of applicants requiring similar waiver consideration.

(2) An important aspect of a waiver request is the recommendations of the District staff. While even the strongest recommendation cannot make a noncompetitive applicant competitive, it does serve to influence significantly the outcome of determinations when an applicant cannot demonstrate overwhelming potential, yet is not sufficiently below standards to warrant disapproval of the request. In all waiver decisions, a detached, objective viewpoint is necessary to ensure success of the primary mission of Navy Recruiting Command.

(3) Either the District CO, XO, EPO, or DIR NAVCRUITPROSTA must conduct a personal interview with the applicant.

2B-3 CNRC Medical Waiver Procedures. Requests for waiver consideration of medical eligibility requirements must be sent to CNRC (Attn: Code 00M). Refer to Section 2J for requirements, standards for waiver consideration, and details concerning the Delayed Entry Medical Program. Forward waiver requests by fax with a CNRC Waiver Cover Letter, Waiver Briefing Sheet and photocopies of the applicant's DD Form 2808 "Report of Medical Examination", DD Form 2807-1 "Report of Medical History", support medical documents (if applicable), surgical reports (if applicable), and applicant's statement concerning the condition (if applicable).

2B-5 Delayed Entry Full Kit Waiver (DEF) Program. The Delayed Entry Full Kit Waiver (DEF) program has been developed to expedite the waiver process by allowing enlistment in the Delayed Entry Program (DEP) based on a District Commanding Officer's Provisional DEF Waiver.

a. Applicants eligible for a District Provisional DEF Waiver have:

(1) No Chart C or D offense criminal history within the past six months (from date of adjudication/final disposition).

(2) No criminal history involving the discharge of a weapon, physical violence, stalking, sexual misconduct, or more than one drug or two behind-the-wheel alcohol convictions.

(3) Applicants must not exceed waiverable limits in any category (i.e. applicants being considered for waiver in accordance with paragraph 2N-2).

(4) HP3 applicants requiring a CNRC waiver must be in "Q" or "W" status only.

b. Prior to contracting, prepare the waiver package according to section 2B-7. Use the Waiver Briefing Sheet and CNRC Waiver Cover Letter to indicate that a district provisional DEF waiver is requested pending the final determination of the waiver. Brief the applicant on the provisional status of the district waiver and have the applicant sign a NAVPERS 1070/613, Delayed Entry Full Kit Waiver Administrative Remarks, Exhibit 2-2.

c. Upon receipt of the waiver package the PRIDE Waiver Section will be annotated to show receipt of the request and authorization for the DEF PRIDE buying option. Once the receipt and authorization are in the PRIDE Waiver Section, the Classifier may buy a DEF PRIDE option. Use of the DEF option will ensure the applicant has a reservation while preventing the conversion of the reservation to the DEP buying option and shipping until receipt of an approved CNRC waiver. The shipping date assigned must be more that 45 days from the date the waiver request is forwarded to CNRC to allow adequate time to review the request.

d. Seats bought through the DEF option will appear on the NETCON.

e. In cases where CNRC disapproves the waiver, the DEF reservation must be cancelled and the applicant must be DEP discharged.

f. Applicants bought through the DEF buying option who have their waiver disapproved must be converted to DEP prior to being cancelled from PRIDE.

2B-6 CNRC Waiver Procedures.

a. Unless otherwise specified in this manual, requests for waiver consideration of enlistment eligibility requirement or program qualification and requests for determination of suitability for enlistment must be sent to Commander, Navy Recruiting Command (CNRC) (Attn: Code 334). In cases where a BUMED waiver and CNRC waiver are required, the BUMED will be run first, then the CNRC waiver using the following procedures:

b. To request a CNRC waiver, forward the applicant's applicable documents required by the Notes annotated on the CNRC Waiver Cover Letter (Exhibit 2-3). Also include any other substantiating documents or pertinent facts not listed in Exhibit 2-3. Securely staple the cover letter to the enclosures. Original documents are not required (photocopies are acceptable).

c. The applicant's preenlistment kit must include a handwritten statement by the applicant on why they want to enlist in the Navy, and where applicable, a handwritten statement providing full details describing each civil conviction or adverse adjudication, lost time or UCMJ conviction, or disqualifying reenlistment code. Handwritten statements on a plain sheet of white paper must include, where applicable:

(1) The receipt of a disqualifying reenlistment code or other than honorable discharge.

(2) Any period(s) of lost time or UCMJ conviction(s) including non-judicial punishment awarded during any previous term(s) of military service.

(3) All civil offenses listed in Charts A, B, C, or D, regardless of disposition (except minor traffic violations that resulted in a fine of $100.00 or less).

(4) All disclosures resulting in a MEPCOM Form 601-23 must have a statement explaining why the disclosure was not listed on DD Form 1966 or the *United States Navy Illicit Behavior Screening Certificate* (NAVCRUIT 1133/65). This statement will be documented in Section VI, Remarks, of DD Form 1966 or the Remarks Section of NAVCRUIT 1133/65.

d. Any statements required by the Electronic Personnel Security Questionnaire (EPSQ). When a statement required by the EPSQ has been made in the appropriate remarks section of the applicant's completed Electronic Personnel Security Questionnaire (EPSQ), a separate handwritten statement is not required.

e. The Navy decides on final acceptability for applicants who have been involved with civil authorities after considering the following factors:

(1) Evaluation and recommendation of the Commanding Officer.

(2) Nature of conviction(s) and degree of participation.

(3) Age at time of conviction(s).

(4) Length of time since last conviction.

(5) Established pattern of conviction.

(6) Record of behavior and attendance at school.

(7) Home environment at time of conviction(s)

(8) Results of home visits, interviews with school officials, probation officers, or other persons who are familiar with the applicant's reputation and standing in the community.

(9) Applicant's motivation toward serving his or her country in the Navy.

f. The Waivers Branch (Code 334) receives requests for waivers submitted to or via CNRC.

g. All recommendations the Waivers Branch makes on requests for waivers are the result of a thorough and careful review by officers assigned to the Division.

h. In each request reviewed, the specific merits or liabilities of a request are carefully weighed. The "whole person" concept is the general rule followed in deliberations, however specific information may take precedence in some cases depending on the nature of the request. An important factor in all deliberations is the relative competitiveness of applicants requiring similar waiver consideration.

i. The single most important aspect of a waiver request is the Commanding Officer's recommendation, particularly in the case of a felony waiver. While even the strongest recommendation cannot make a noncompetitive applicant competitive, it does serve to influence significantly the outcome of determinations when an applicant cannot demonstrate overwhelming potential, yet is not sufficiently below standards to warrant disapproval of the request. Specific recommendations are desired. An endorsement merely forwarding a request will not be considered favorably. In all decisions for waiver of eligibility requirements or program qualifications, a detached, objective viewpoint is necessary to ensure success of the primary mission of Navy Recruiting Command.

Marine Corps

The following is from MCO P1100.72B.

2. Waiver Recommendations. Waivers will be recommended for only two reasons:

a. Highly favorable traits or mitigating circumstances exist which outweigh the reason for disqualification; or

b. The enlistment/reenlistment is clearly in the best interests of the Marine Corps.

3. Waiver Responsibilities. The responsibility of determining whether or not waiver requests warrant favorable consideration rests with all levels of command, but initially with the recruiter. The recruiter's responsibilities include:

a. Screening (questioning and counseling),

b. Investigating (gathering proper documentation), and

c. Initiating the waiver request, if warranted.

3301. ELIGIBILITY FOR WAIVERS.

1. Making a Determination. To determine whether an applicant is eligible for a waiver, the following matters must be considered:

a. Is the enlistment criteria/standard one which may or may not be waived? Refer to Table 3-16, page 3-167, for ineligible conditions which are not waiverable.

b. SUBORDINATE COMMANDERS ARE NOT AUTHORIZED TO IMPOSE STRICTER CRITERIA TO OFFICIAL WAIVER GUIDELINES. HOWEVER, COMMANDERS AT ANY LEVEL MAY DISAPPROVE A WAIVER BASED UPON THE WAIVER'S OWN MERIT, THE COMMANDER'S JUDGMENT, AND THE CURRENT RECRUITING ENVIRONMENT.

c. An applicant's eligibility will be based on the level of education, the mental category, and on the "whole person" concept.

2. The "Whole Person" Concept.

a. Waivers will be evaluated using the "whole person" concept. Under this concept, an applicant's qualifications are compared with past performance with the intent of calculating potential effectiveness in the Marine Corps. Such an evaluation is difficult. The evaluation should present for consideration all relevant facts and information, as well as a thorough meaningful evaluation. Waiver requests which simply identify the disqualifying factor(s) without thorough discussion of all mitigating circumstances and the applicant's favorable traits are a disservice to the applicant and may well jeopardize waiver approval.

b. To help in evaluating cases, tally strengths and weaknesses. Be alert for patterns of success or failure. Ask yourself the following questions:

(1) Is the applicant a desirable prospect?

(2) Does the applicant's strengths outweigh the reason(s) for disqualification?

(3) Are the applicant's demonstrated qualities indicative of successful service as a Marine?

(4) Is the applicant's enlistment/reenlistment clearly in the best interest of the Marine Corps?

c. If there is any doubt, or the answer to any of the above questions is "no," a request for a waiver should not be processed. This decision must be made without regard to monthly production goals.

3302. WAIVER AUTHORITY LEVELS. Requests for waiver of enlistment/ reenlistment criteria will be submitted to the appropriate waiver authority depicted in table 3-19, page 3-179, for decision.

3304. SUBMISSION OF WAIVERS.

1. Waiver requests will be forwarded via the chain of command. Each level of command must act on the request and provide an appropriate endorsement. However, recruiting station commanding officers, district commanding officers, and commanding officers, and commanding generals of the recruiting regions have authority to disapprove the enlistment or reenlistment of an applicant without forwarding the case to the next higher level.

a. Medical and prior-service (PSEP/FTAP) waiver requests may be submitted directly to the CG MCRC from the recruiting station unless there are other disqualifying factors (e.g., dependent, moral, etc.) which fall within the waiver authority of the district commanding officer or region commanding general, or unless stricter conditions are imposed by those commanders.

Estimates of Prevalence of Health Conditions, by Race and Ethnicity

Table C.1
Prevalence of Specific Health Conditions Among Males Ages 18–25, by Race/Ethnicity (%)

Condition	White	Black	Mexican	Hispanic
Major				
Functional limitation	5.44	5.02	3.45	3.92
Asthma in past year	3.17	3.55	1.49	2.03
Ulcer	3.38	1.06	1.44	1.95
Heart condition	2.92	3.04	1.27	1.32
Hypertension	1.88	4.00	0.26	1.03
Need special equipment	0.94	1.17	0.29	0.28
Diabetes	0.90	0.67	0.94	0.93
Organ failure	0.53	0.29	0.60	0.89
Blind/deaf	0.30	0.81	0.00	0.00
Angina	0.12	0.00	0.77	0.52
Chronic heart disease	0.11	0.15	0.00	0.07
Stroke	0.11	0.00	0.07	0.05
Myocardial infarction	0.10	0.05	0.06	0.25
Emphysema	0.04	0.12	0.10	0.07
Minor				
Sinusitis	8.72	9.21	3.15	4.12
Hay fever	7.30	6.17	2.12	3.18
Chronic bronchitis	2.96	2.52	1.26	1.67
Cancer	0.91	0.00	0.54	0.39

NOTE: Unadjusted estimates.

Table C.2
Prevalence of Specific Health Conditions Among Females Ages 18–25, by Race/Ethnicity (%)

Condition	White	Black	Mexican	Hispanic
Major				
Functional limitation	9.51	9.01	4.98	3.92
Asthma in past year	6.43	4.68	2.56	3.06
Ulcer	4.88	2.23	2.19	1.95
Heart condition	4.12	2.67	1.61	1.32
Hypertension	2.53	4.08	2.57	1.03
Need special equipment	0.86	0.82	0.17	0.28
Diabetes	1.91	2.22	0.94	0.93
Organ failure	1.50	1.06	0.78	0.89
Blind/deaf	0.09	0.02	0.23	0.00
Angina	0.09	0.18	0.42	0.52
Chronic heart disease	0.07	0.18	0.00	0.07
Stroke	0.04	0.40	0.12	0.05
Myocardial infarction	0.07	0.18	0.07	0.25
Emphysema	0.06	0.18	0.06	0.07
Minor				
Sinusitis	15.83	11.51	3.58	5.26
Hay fever	8.63	5.88	2.61	3.53
Chronic bronchitis	5.65	2.79	1.51	2.46
Cancer	1.72	0.98	0.48	0.63

NOTE: Unadjusted estimates.

Table C.3
Prevalence of Specific Health Conditions Among Males Ages 14–17, by Race/Ethnicity (%)

Condition	White	Black	Puerto Rican	Mexican	Cuban	Other Hispanic	All Hispanic
Major							
Asthma (ever)	15.92	18.26	35.86	10.91	20.97	14.62	14.76
Functional limitation	11.12	13.78	19.04	4.71	9.90	8.00	7.13
Other developmental problem	3.83	3.75	5.14	1.24	4.59	4.23	2.37
Mobility impairment	2.66	2.55	5.39	1.86	0.00	2.74	2.36
Blind/deaf	2.65	3.37	2.27	3.52	9.44	2.54	3.39
Heart condition	1.45	1.42	1.77	0.76	4.53	1.19	1.09
Need special equipment	1.13	1.59	0.95	2.15	0.00	0.89	1.69
Cerebral palsy	0.58	0.32	2.28	0.53	1.74	0.62	0.79
Diabetes	0.45	0.22	0.00	0.18	0.00	0.00	0.12
Autism	0.37	0.12	1.38	0.03	4.59	0.00	0.33
Congenital heart disease	0.18	0.07	0.00	0.21	0.00	0.68	0.27
Juvenile arthritis	0.17	0.05	0.00	0.09	0.00	0.00	0.06
Down's syndrome	0.12	0.00	0.46	0.08	0.00	0.32	0.17
Muscular dystrophy	0.03	0.00	0.00	0.33	0.00	0.00	0.22
Cystic fibrosis	0.00	0.00	0.00	0.00	0.00	0.00	0.00
Sickle cell anemia	0.00	0.56	0.00	0.00	0.00	0.00	0.00
Minor							
Hay fever	18.21	10.74	16.35	7.18	6.56	14.63	9.64
Takes Rx meds > 3 months	17.27	10.34	17.09	5.81	10.17	11.97	8.43
Respiratory allergies	15.37	10.34	20.04	6.41	4.23	9.80	8.52
ADD	13.59	8.42	13.95	3.97	11.62	7.71	6.08
Frequent headaches	7.51	8.16	9.97	5.22	6.81	8.32	6.41
Skin allergies	7.16	6.88	11.23	5.58	0.00	7.71	6.44
Digestive allergies	3.77	1.53	4.43	2.21	0.88	2.51	2.47
Seizure disorder	0.45	1.00	1.99	0.44	0.00	0.95	8.52

Table C.4
Prevalence of Specific Health Conditions Among Females Ages 14–17, by Race/Ethnicity (%)

Condition	White	Black	Puerto Rican	Mexican	Cuban	Other Hispanic	All Hispanic
Major							
Asthma (ever)	12.13	13.55	32.13	12.02	8.93	10.46	13.96
Functional limitation	6.46	6.43	9.67	3.74	7.58	3.05	4.38
Other developmental problem	2.60	2.08	4.01	1.95	0.61	0.44	1.81
Mobility impairment	3.59	2.36	5.86	2.25	0.61	1.47	2.45
Blind/deaf	3.07	5.02	4.60	4.15	0.00	7.36	4.84
Heart condition	1.29	0.54	0.00	0.78	0.00	1.71	0.88
Need special equipment	1.47	0.40	1.99	0.71	0.00	0.47	0.79
Cerebral palsy	0.45	0.02	1.99	0.40	0.79	0.18	0.54
Diabetes	0.31	0.00	0.00	0.00	0.00	0.00	0.00
Autism	0.12	0.25	0.00	0.00	0.00	0.00	0.00
Congenital heart disease	0.29	0.02	0.00	0.15	0.00	0.19	0.14
Juvenile arthritis	0.29	0.14	0.00	0.05	0.00	0.00	0.03
Down's syndrome	0.05	0.04	0.36	0.00	0.00	0.00	0.04
Muscular dystrophy	0.06	0.00	0.00	0.00	0.00	0.00	0.00
Cystic fibrosis	0.01	0.00	0.00	0.00	0.00	0.00	0.00
Sickle cell anemia	0.00	0.53	0.00	0.25	0.00	0.57	0.29
Minor							
Hay fever	14.90	13.45	11.38	7.46	2.14	9.32	8.22
Takes Rx meds > 3 months	17.68	9.05	10.08	8.31	6.07	5.42	7.80
Respiratory allergies	15.84	11.50	16.69	9.94	5.93	8.30	10.26
ADD	4.95	2.06	1.73	2.82	11.34	3.34	3.03
Frequent headaches	13.07	9.99	12.06	13.02	5.83	10.97	12.25
Skin allergies	7.73	6.52	10.96	4.75	0.00	6.03	5.66
Digestive allergies	4.25	3.41	2.01	3.28	4.09	2.72	3.02
Seizure disorder	0.91	0.79	1.99	0.70	0.61	0.06	0.71

Tests of Statistical Significance in Medical Disqualification Rates

Table D.1
P-Values from Test of Difference in Proportion Failing New Accession Weight Standards Between Successive Restriction Categories, Adult Male Sample

	U.S. Citizen	High School Graduate
White	0.495	0.002
Black	0.134	0.232
Mexican	0.019	0.621
All Hispanic	0.001	0.475

Table D.2
P-Values from Test of Difference in Proportion Failing Air Force Weight Standards Between Successive Restriction Categories, Adult Male Sample

	U.S. Citizen	High School Graduate
White	0.557	0.000
Black	0.001	0.086
Mexican	0.000	0.616
All Hispanic	0.000	0.817

Table D.3
P-Values from Test of Difference in Proportion with a Major Condition Between Successive Restriction Categories, Adult Male Sample

	U.S. Citizen	High School Graduate
White	0.000	0.000
Black	0.000	0.000
Mexican	0.004	0.005
All Hispanic	0.000	0.044

Table D.4
P-Values from Test of Difference in Proportion with a Nonwaiverable Condition (Major or Weight) Between Successive Restriction Categories, Adult Male Sample

	U.S. Citizen	High School Graduate
White	0.263	0.026
Black	0.000	0.152
Mexican	0.043	0.882
All Hispanic	0.000	0.000

Table D.5
P-Values from Test of Difference in Proportion with at Least One Disqualifying Condition (Major, Minor, or Weight) Between Successive Restriction Categories, Adult Male Sample

	U.S. Citizen	High School Graduate
White	0.204	0.425
Black	0.001	0.033
Mexican	0.072	0.831
All Hispanic	0.001	0.000

**Table D.6
P-Values from Test of Difference in Proportion
Failing New Accession Weight Standards
Between Successive Restriction Categories,
Adult Female Sample**

	U.S. Citizen	High School Graduate
White	0.456	0.000
Black	0.000	0.000
Mexican	0.363	0.000
All Hispanic	0.170	0.431

**Table D.7
P-Values from Test of Difference in Proportion
Failing Air Force Weight Standards Between
Successive Restriction Categories,
Adult Female Sample**

	U.S. Citizen	High School Graduate
White	0.020	0.000
Black	0.000	0.625
Mexican	0.080	0.538
All Hispanic	0.061	0.007

**Table D.8
P-Values from Test of Difference in Proportion
with a Major Condition Among Successive
Restriction Categories, Adult Female Sample**

	U.S. Citizen	High School Graduate
White	0.000	0.000
Black	0.039	0.000
Mexican	0.350	0.904
All Hispanic	0.113	0.110

Table D.9
P-Values from Test of Difference in Proportion with a Nonwaiverable Condition (Major or Weight) Between Successive Restriction Categories, Adult Female Sample

	U.S. Citizen	High School Graduate
White	0.452	0.001
Black	0.000	0.000
Mexican	0.043	0.000
All Hispanic	0.000	0.025

Table D.10
P-Values from Test of Difference in Proportion with at Least One Disqualifying Condition (Major, Minor or Weight) Between Successive Restriction Categories, Adult Female Sample

	U.S. Citizen	High School Graduate
White	0.590	0.026
Black	0.016	0.000
Mexican	0.000	0.000
All Hispanic	0.000	0.097

Table D.11
P-Values from Test of Difference in Proportion Failing New Accession Weight Standards Between Race/Ethnic Categories, Adult Male Sample

	Ages 18–25	U.S. Citizen	High School Graduate
Black	0.000	0.000	0.001
Mexican	0.031	0.048	0.093
All Hispanic	0.075	0.012	0.057

Table D.12
P-Values from Test of Difference in Proportion Failing Air Force Weight Standards Between Race/Ethnic Categories, Adult Male Sample

	Ages 18–25	U.S. Citizen	High School Graduate
Black	0.000	0.000	0.000
Mexican	0.004	0.000	0.013
All Hispanic	0.044	0.001	0.042

Table D.13
P-Values from Test of Difference in Proportion with a Major Condition Between Race/Ethnic Categories, Adult Male Sample

	Ages 18–25	U.S. Citizen	High School Graduate
Black	0.868	0.030	0.000
Mexican	0.000	0.280	0.288
All Hispanic	0.001	0.461	0.000

Table D.14
P-Values from Test of Difference in Proportion with a Nonwaiverable Condition (Major or Weight) Between Race/Ethnic Categories, Adult Male Sample

	Ages 18–25	U.S. Citizen	High School Graduate
Black	0.000	0.000	0.000
Mexican	0.653	0.384	0.489
All Hispanic	0.666	0.383	0.583

Table D.15
P-Values from Test of Difference in Proportion with at Least One Disqualifying Condition (Major, Minor or Weight) Between Race/Ethnic Categories, Adult Male Sample

	Ages 18–25	U.S. Citizen	High School Graduate
Black	0.000	0.000	0.000
Mexican	0.047	0.815	0.862
All Hispanic	0.081	0.981	0.760

Table D.16
P-Values from Test of Difference in Proportion Failing New Accession Weight Standards Between Race/Ethnic Categories, Adult Female Sample

	Ages 18–25	U.S. Citizen	High School Graduate
Black	0.000	0.000	0.000
Mexican	0.000	0.000	0.000
All Hispanic	0.000	0.014	0.000

Table D.17
P-Values from Test of Difference in Proportion Failing Air Force Weight Standards Between Race/Ethnic Categories, Adult Female Sample

	Ages 18–25	U.S. Citizen	High School Graduate
Black	0.000	0.000	0.000
Mexican	0.000	0.000	0.000
All Hispanic	0.000	0.000	0.000

Table D.18
P-Values from Test of Difference in Proportion with a Major Condition Between Race/Ethnic Categories, Adult Female Sample

	Ages 18–25	U.S. Citizen	High School Graduate
Black	0.000	0.000	0.000
Mexican	0.000	0.129	0.368
All Hispanic	0.000	0.174	0.308

Table D.19
P-Values from Test of Difference in Proportion with a Nonwaiverable Condition (Major or Weight) Among Race/Ethnic Categories, Adult Female Sample

	Ages 18–25	U.S. Citizen	High School Graduate
Black	0.000	0.000	0.000
Mexican	0.298	0.056	0.131
All Hispanic	0.959	0.189	0.228

Table D.20
P-Values from Test of Difference in Proportion with at Least One Disqualifying Condition (Major, Minor or Weight) Among Race/Ethnic Categories, Adult Female Sample

	Ages 18–25	U.S. Citizen	High School Graduate
Black	0.000	0.000	0.028
Mexican	0.004	0.757	0.596
All Hispanic	0.026	0.564	0.335

References

Achatz, Mary, Martha Stapleton Kudela, Shelley Perry, and Jerome D. Lehnus, "Career Plans and Military Propensity of Hispanic Youth: Interviews with 1997 YATS Respondents," Arlington, Va.: Defense Manpower Data Center, 2000.

AETC Instruction 36-2002—*See* U.S. Air Force Air Education and Training Command, 2003.

AR 40-501—*See* Headquarters Department of the Army, 2002.

AR 601-210—*See* Headquarters Department of the Army, 1995.

AR 601-280—*See* Headquarters Department of the Army, 1999.

Asch, Beth J., Christopher Buck, Jacob Alex Klerman, Meredith Kleykamp, and David Loughran, *What Factors Affect the Military Enlistment of Hispanic Youth? A Look at Enlistment Qualifications*, Santa Monica, Calif.: RAND Corporation, DB-484-OSD, 2005. As of August 25, 2008:
http://www.rand.org/pubs/documented_briefings/DB484/

Asch, Beth J., and James Hosek, *Looking to the Future: What Does Transformation Mean for Military Manpower and Personnel Policy?* Santa Monica, Calif.: RAND Corporation, OP-108-OSD, 2004. As of August 25, 2008:
http://www.rand.org/pubs/occasional_papers/OP108/

Asch, Beth J., M. Rebecca Kilburn, and Jacob Alex Klerman, *Attracting College-Bound Youth into the Military: Toward the Development of New Recruiting Policy Options,* Santa Monica, Calif.: RAND Corporation, MR-984-OSD, 1999. As of August 25, 2008:
http://www.rand.org/pubs/monograph_reports/MR984/

Asch, Beth J., and David S. Loughran, *Reserve Recruiting and the College Market: Is a New Educational Benefit Needed?* Santa Monica, Calif.: RAND Corporation, TR-127-OSD, 2005. As of August 25, 2008:
http://www.rand.org/pubs/technical_reports/TR127/

Asch, Beth J., John A. Romley, and Mark E. Totten, *The Quality of Personnel in the Enlisted Ranks,* Santa Monica, Calif.: RAND Corporation, MG-324-OSD, 2005. As of August 25, 2008:
http://www.rand.org/pubs/monographs/MG324/

Asch, Beth J., and John T. Warner, *A Theory of Military Compensation and Personnel Policy,* Santa Monica, Calif.: RAND Corporation, MR-439-OSD, 1994. As of August 25, 2008:
http://www.rand.org/pubs/monograph_reports/MR439/

Bachman, J. G., P. Freedman-Doan, and P. M. O'Malley, "Should U.S. Military Recruiters Write Off the College-Bound?" *Armed Forces & Society,* Vol. 27, No. 3, 2001, pp. 461–476.

Badillo, Gilbert, and G. David Curry, "The Social Incidence of Vietnam Casualties: Social Class or Race?" *Armed Forces & Society,* Vol. 2, No. 3, 1976, pp. 387–406.

Binkin, Martin, and Mark Eitelberg, with Alvin Schexnider and Marvin Smith, *Blacks and the Military,* Washington, D.C.: The Brookings Institution, Studies in Defense Policy, 1982.

Boehmer, Matt, and Andrea Zucker, *Department of Defense Youth Poll Wave 6—November 2003: Overview Report,* Arlington, Va.: Department of Defense, Defense Human Resources Activity, Joint Advertising, Market Research and Studies, 2004.

Boes, Jennifer O'Connor, Martin F. Wiskoff, and Marc Flacks, "Hispanic Youth and Military Enlistment Propensity," Monterey, Calif.: Security Research Center Defense Security Service, 1999.

Buddin, Richard J., *Success of First-Term Soldiers: The Effects of Recruiting Practices and Recruit Characteristics*, Santa Monica, Calif.: RAND Corporation, MG-262, 2005. As of August 25, 2008:
http://www.rand.org/pubs/monographs/MG262/

Buddin, Richard J., Daniel S. Levy, Janet M. Hanley, and Donald M. Waldman, *Promotion Tempo and Enlisted Retention*, Santa Monica, Calif.: RAND Corporation, R-4135-FMP, 1992. As of August 25, 2008:
http://www.rand.org/pubs/reports/R4135/

Cameron, Stephen V., and James J. Heckman, "The Dynamics of Educational Attainment for Black, Hispanic, and White Males," *Journal of Political Economy*, Vol. 109, No. 3, 2001, pp. 455–499.

Carter-Pokras, Olivia D., and Peter J. Gergen, "Reported Asthma Among Puerto Rican, Mexican-American, and Cuban Children, 1982 Through 1984," *American Journal of Public Health*, Vol. 83, 1993, pp. 580–582.

Cho, Y., and R. A. Hummer, "Disability Status Differentials across Fifteen Asian and Pacific Islander Groups and the Effect of Nativity and Duration of Residence in the U.S.," *Social Biology*, Vol. 48, No. 3–4, 2001, pp. 171–195.

COMNAVCRUITCOMINST 1130.8F—*See* U.S.Navy Recruiting Command, 2003.

Defense Human Resources Activity, Joint Advertising Market Research, December 2008 Youth Poll 14, Findings Presentation. As of April 4, 2008:
http://www.dmren.org/DMREN/execute/secure/documents/youth (password-protected)

DeNavas-Walt, Carmen, Robert Cleveland, and Bruce H. Webster, Jr., "Income in the United States: 2002," U.S. Census Bureau Current Population Reports, Washington, D.C.: U.S. Government Printing Office, 2003.

Department of Defense, Recruit Market Information System, undated. As of September 5, 2006:
https://www.dmdc.osd.mil/rmis (password-protected)

Department of Defense, *Qualification Standards for Enlistment, Appointment, and Induction*, Washington, D.C.: DoD Directive 1304.26, March 4, 1994.

———, *Population Representation in the Military Services, Fiscal Year 1997*, Washington, D.C.: Office of the Under Secretary of Defense for Personnel and Readiness. As of April 4, 2008:
http://www.defenselink.mil/prhome/poprep97/

———, *Criteria and Procedure Requirements for Physical Standards for Appointment, Enlistment, or Induction in the Armed Forces*, Washington, D.C.: Instruction Number 6130.4, 2000.

———, *Selected Military Compensation Tables: January 1, 2001*, Arlington, Va.: Directorate of Compensation, Department of Defense, 2001.

———, *Medical Processing and Examinations*, Washington, D.C.: United States Military Entrance Processing Command, USMEPCOM Regulation 40-1, 2002.

———, *Population Representation in the Military Services, Fiscal Year 2002*, Washington, D.C.: Office of the Under Secretary of Defense for Personnel and Readiness, 2004. As of July 12, 2005:
http://www.defenselink.mil/prhome/poprep2002

———, *Population Representation in the Military Services Fiscal Year 2004*, Washington, D.C.: Office of the Under Secretary of Defense for Personnel and Readiness, 2006.

DoD Directive 1304.26—*See* Department of Defense, 1994.

Duncan, Brian, V. Joseph Hotz, and Stephen J. Trejo, "Hispanics in the U.S. Labor Market," in Marta Tienda and Faith Mitchell, eds., *Hispanics and the Future of America*, Washington, D.C.: National Research Council of the National Academies Press, Committee on Population, Division of Behavioral and Social Sciences and Education, Panel on Hispanics in the United States, 2006, pp. 228–290.

Franzini, L., J. C. Ribble, and A. M. Keddie, "Understanding the Hispanic Paradox," *Ethnicity & Disease*, Vol. 11, No. 3, 2001, pp. 496–518.

Freid, V. M., K. Prager, A. P. MacKay, and H. Xia, *Chartbook on Trends in the Health of Americans*, Hyattsville, Md.: National Center for Health Statistics, 2003.

Frisbie, W. P., Y. Cho, and R. A. Hummer, "Immigration and the Health of Asian and Pacific Islander Adults in the United States," *American Journal of Epidemiology*, Vol. 153, No. 4, 2001, pp. 372–380.

Fry, Richard, *Hispanic Youth Dropping Out of U.S. Schools: Measuring the Challenge*, Washington, D.C.: Pew Hispanic Center, 2003.

Ganderton, Philip T., and Richard Santos, "Hispanic College Attendance and Completion: Evidence from the High School and Beyond Surveys," *Economics of Education Review,* Vol. 14, No. 1, 1995, pp. 35–46.

Gimbel, Cynthia, and Alan Booth, "Who Fought in Vietnam?" *Social Forces*, Vol. 74, No. 4, 1996, pp. 1137–1157.

Gotz, Glenn, and John McCall, *A Dynamic Retention Model for Air Force Officers: Theory and Estimates,* Santa Monica, Calif.: RAND Corporation, R-3028-AF, 1984. As of August 29, 2008:
http://www.rand.org/pubs/reports/R3028/

Hattiangadi, Anita U., Gary Lee, and Aline O. Quester, *Recruiting Hispanics: The Marine Corps Experience, Final Report*, Alexandria, Va.: The CNA Corporation, CRM D0009071.A2/Final, 2004.

Hattiangadi, Anita U., Aline O. Quester, SgtMaj (Ret) Gary Lee, Diana Lien, and Ian MacLeod, *Non-Citizens in Today's Military: Final Report*, Alexandria, Va.: The CNA Corporation, CRM D0011092.A2/Final, 2005.

Headquarters Department of the Army, *Regular Army and Army Reserve Enlistment Program,* Washington, D.C.: Army Regulation (AR) 601-210, March 28, 1995.

———, *Army Retention Program*, Washington, D.C.: AR 601-280, 1999.

———, *Standards of Medical Fitness*, Washington, D.C.: AR 40-501, September 30, 2002.

———, *The Army Weight Control Program,* Washington, D.C.: AR 600-9, November 27, 2006.

Headquarters U.S. Marine Corps, *Military Personnel Procurement Manual,* Volume 2, *Enlisted Procurement,* Washington, D.C.: MCO P1100.72B, 1997.

———, *Military Personnel Procurement Manual,* Volume 2, *Enlisted Procurement,* Washington, D.C.: MCO P1100.72C, 2004.

Heckman, James, "Lessons from the Bell Curve," *Journal of Political Economy*, Vol. 103, No. 5, 1995, pp. 1091–1120.

Heckman, James, and Edward Vytlacil, "Identifying the Role of Cognitive Ability in Explaining the Level of and Change in the Return to Schooling," *The Review of Economics and Statistics*, Vol. 83, No. 1, 2001, pp. 1–12.

Herrnstein, Richard, and Charles Murray, *The Bell Curve: Class Structure in American Life*, New York: The Free Press, 1994.

Hoffman, Kathryn, Charmaine Llagas, and Thomas D. Snyder, *Status and Trends in the Education of Blacks*, U.S. Department of Education, National Center for Education Statistics, NCES 2003–034, 2003. As of August 25, 2008:
http://nces.ed.gov/Pubsearch/pubsinfo.asp?pubid=2003034

Hosek, James, and Michael Mattock, *Learning About Quality: How the Quality of Military Personnel Is Revealed over Time*, Santa Monica, Calif.: RAND Corporation, MR-1593-OSD, 2003. As of August 25, 2008:
http://www.rand.org/pubs/monograph_reports/MR1593/

Hotz, V. Joseph, Lixin Colin Xu, Marta Tienda, and Avner Ahituv, "Are There Returns to the Wages of Young Men from Working While in School?" *Review of Economics and Statistics*, Vol. 84, No. 2, May 2002, pp. 222–236.

Human Resources Research Organization, homepage. As of October 30, 2008:
http://www.humrro.org/corpsite/

Karoly, Lynn A., and Jacob A. Klerman, Young Men and the Transition to Stable Employment, *Monthly Labor Review*, Vol. 117, No. 8, 1994, pp. 31–48.

Kilburn, M. Rebecca, "Minority Representation in the U.S. Military," dissertation, Chicago: University of Chicago Department of Economics, 1992.

Kilburn, M. Rebecca, and Beth J. Asch, *Recruiting Youth in the College Market: Current Practices and Future Policy Options,* Santa Monica, Calif.: RAND Corporation, MR-1093-OSD, 2003. As of August 25,2008: http://www.rand.org/pubs/monograph_reports/MR1093/

Kilburn, M. Rebecca, and Jacob Alex Klerman, *Enlistment Decisions in the 1990s: Evidence from Individual-Level Data,* Santa Monica, Calif.: RAND Corporation, MR-944-OSD, 1999. As of August 25, 2008: http://www.rand.org/pubs/monograph_reports/MR944/

Knouse, S. B., "Social Support for Hispanics in the Military," *International Journal of Intercultural Relations,* Vol. 15, 1991, pp. 427–444.

Laurence, H. Janice, Peter F. Ramsberger, and Jane Arabian, *Education Credential Tier Evaluation*, Alexandria, Va.: Human Resources Research Organization, FR-EADD-96-19, 1996.

Leal, David L., "The Multicultural Military: Military Service and the Acculturation of Latinos and Anglos," *Armed Forces & Society,* Vol. 29, No. 2, 2003, pp. 205–226.

Lindquist, C. H., and R. M. Bray, "Trends in Overweight and Physical Activity among U.S. Military Personnel, 1995–1998," *Preventive Medicine,* Vol. 32, No. 2, 2001, pp. 57–65.

Llagas, Charmaine, "Status and Trends in the Education of Hispanics," Washington, D.C.: U.S. Department of Education National Center for Education Statistics, 2003.

MCO P1100.72B—*See* Headquarters U.S. Marine Corps, 1997.

MCO P1100.72C—*See* Headquarters U.S. Marine Corps, 2004.

Moskos, C. C., and J. S. Butler, *All That We Can Be: Black Leadership and Racial Integration the Army Way,* New York: Twentieth Century Fund, 1996.

Neal, Derek A., and William R. Johnson, "The Role of Premarket Factors in Black-White Wage Differences," *Journal of Political Economy,* Vol. 104, No. 5, 1996, pp. 869–895.

Neumark, David, "Youth Labor Markets in the United States: Shopping Around vs. Staying Put," *The Review of Economics and Statistics,* Vol. 84, No. 3, 2002, pp. 462–482.

NLSY97 User's Guide, Columbus, Ohio: Center for Human Resource Research, 2005.

Nolte, R., S. C. Franckowiak, C. J. Crespo, and R. E. Andersen, "U.S. Military Weight Standards: What Percentage of U.S. Young Adults Meet the Current Standards?" *American Journal of Medicine,* Vol. 113, No. 6, 2002, pp. 486–490.

Ogden, Cynthia L., Katherine M. Flegal, Margaret D. Carroll, and Clifford I. Johnson, "Prevalence and Trends in Overweight Among U.S. Children and Adolescents, 1999–2000," *Journal of the American Medical Association,* Vol. 288, No. 14, 2002, pp. 1728–1732.

Palloni, A., and J. D. Morenoff, "Interpreting the Paradoxical in the Hispanic Paradox: Demographic and Epidemiologic Approaches," *Annals of the New York Academy of Science,* Vol. 954, 2001, pp. 140–174.

Pellum, Maj Martin W., USAF, *Air Force Recruiting: Considerations for Increasing the Proportion of Black and Hispanic Persons in the Enlisted Force,* Maxwell Air Force Base, Ala.: Air University Press, 1996.

Pew Hispanic Center, "Hispanics in the Military," Pew Hispanic Center Fact Sheet, March 27, 2003, Washington, D.C.: Pew Hispanic Center, 2003.

Popkin, Barry M., and J. Richard Udry, "Adolescent Obesity Increases Significantly in Second and Third Generation U.S. Immigrants: The National Longitudinal Study of Adolescent Health," *The Journal of Nutrition,* Vol. 128, No. 4, April 1998, pp. 701–706.

Powers, Rod, "Computing the ASVAB Score: How the Overall ASVAB Score Is Computed," About.com. As of March 19, 2004: http://usmilitary.about.com/library/milinfo/blafqtscore.htm

———, "United States Military Enlistment Standards: Do You Qualify to Enlist in the United States Military?" About.com. As of March 19, 2004:
http://usmilitary.about.com/library/weekly/aa082701a.htm

Quester, Aline O., and Curtis L. Gilroy, "Women and Minorities in America's Volunteer Army," *Contemporary Economic Policy,* Vol. 20, No. 2, April 2002, pp. 111–121.

Report of the President's Commission on an All-Volunteer Force (Gates Commission Report), February 1970.

Rosenfeld, Paul, and Amy L. Culbertson, "Hispanics in the Military," in Amy L. Culbertson, ed., *Hispanics in the Workplace,* Newbury Park, Calif.: Sage Publications, 1992.

Segal, David R., Jerald G. Bachman, Peter Freedman-Doan, and Patrick M. O'Malley, "Propensity to Serve in the U.S. Military: Temporal Trends and Subgroup Differences," *Armed Forces & Society,* Vol. 25, No. 3, 1999, pp. 407–427.

Sellman, W. Steven, "Predicting Readiness for Military Service: How Enlistment Standards Are Established," commissioned paper for the National Assessment Governing Board, 2004. As of July 11, 2005:
http://www.nagb.org/release/sellman.doc

———, "Public Policy Implications for Military Entrance Standards," keynote address presented at the 39th Annual Conference of the International Military Testing Association, Sydney, Australia, October 14–16, 1997.

Senate Committee on Armed Services, Report 93-884, 1974.

Shapiro, David, "Wage Differentials Among Black, Hispanic, and White Young Men," *Industrial and Labor Relations Review,* Vol. 37, No. 4, 1984, pp. 570–581.

Shin, Hyon B., and Rosalind Bruno, "Language Use and English-Speaking Ability: 2000," Washington, D.C.: U.S. Department of Commerce Economics and Statistics Administration, U.S. Census Bureau, 2003.

Smith, D. Alton, Stephen Sylwester, and Christine Villa, "Army Reenlistment Models," in Curtis L. Gilroy, David K. Horne, and D. Alton Smith, eds., *Military Compensation and Personnel Retention: Models and Evidence,* Alexandria, Va.: U.S. Army Research Institute for the Behavioral and Social Sciences, 1991, pp. 43–179.

Smith, James P., "Assimilation Across the Latino Generations," *American Economic Review,* Vol. 3, No. 3, 1993, pp. 315–319.

Swail, Watson Scott, Alberto F. Cabrera, and Chul Lee, *Latino Youth and the Pathway to College,* Washington, D.C.: Pew Hispanic Center, 2004.

Swanson, Christopher B., *Who Graduates? Who Doesn't? A Statistical Portrait of Public High School Graduation, Class of 2001,* Washington, D.C.: The Urban Institute Education Policy Center, 2005.

Taber, Christopher, "The Rising College Premium in the Eighties: Return to College or Return to Unobserved Ability?" *Review of Economic Studies,* Vol. 68, No. 3, 2001, pp. 665–691.

Tienda, Marta, and Faith Mitchell, "Introduction: E Pluribus Plures or E Pluribus Unum," in Marta Tienda and Faith Mitchell, eds., *Hispanics and the Future of America,* Washington D.C.: National Academies Press, Committee on Population, Division of Behavioral and Social Sciences and Education, Panel on Hispanics in the United States, 2006, pp. 1–15.

Triandis, Harry C., "Hispanic Concerns About the U.S. Navy," Champaign, Ill.: University of Illinois, Department of Psychology, 1981.

Trybula, David C., "Three Essays on the Economics of Military Manpower," doctoral dissertation, Austin, Tex.: University of Texas at Austin, 1999.

U.S. Air Force Air Education and Training Command, *Recruiting Procedures for the Air Force,* Washington, D.C.: AETC Instruction 36-2002, March 18, 2003.

U.S. Navy Recruiting Command, *Navy Recruiting Manual–Enlisted,* Washington, D.C.: COMNAVCRUITCOMINST 1130.8F, 2003.

Vega, W. A., and H. Amaro, "Latino Outlook: Good Health, Uncertain Prognosis," *Annual Review of Public Health,* 1994, pp. 39–67.

Wallach, John, "Covenant Training—Great Lakes Special Programs Invest in Recruits, Students." As of September 5, 2008:
http://findarticles.com/p/articles/mi_pnav/is_200204/ai_2688817875

Warner, John, and Gary Solon, "First Term Attrition and Reenlistment in the U.S. Army," in Curtis L. Gilroy, David K. Horne, and D. Alton Smith, eds., *Military Compensation and Personnel Retention: Models and Evidence*, Alexandria, Va.: U.S. Army Research Institute for the Behavioral and Social Sciences, 1991, pp. 243–280.

Wilson, Michael J., James B. Greenlees, Tracey Hagerty, Cynthia V. Helba, D. Wayne Hintze, and Jerome D. Lehnus, "Youth Attitude Tracking Study 1999 Propensity and Advertising Report," Arlington, Va.: Defense Manpower Data Center, 2000.